GROWER'S GUIDES

Jean-Martin Fortier

FROM THE MARKET GARDENER

Fruiting Vegetables

A Grower's Guide

TRANSLATED BY LAURIE BENNETT
EDITED BY PIERRE NESSMANN
WITH THE COLLABORATION OF MATHILDE LEBECQ
AND DAMIEN TERRAL
ILLUSTRATIONS BY FLORE AVRAM

Translated by Laurie Bennett. Cover design by Diane McIntosh.
Printed in Canada. First printing January, 2026.

First published in France
under the title: *Les légumes fruits. Les guides du jardinier-maraîcher,*
Jean-Martin Fortier, Flore Avram.

Any other inquiries can be directed by mail to:
New Society Publishers P.O. Box 189, Gabriola Island, BC
V0R 1X0, Canada

New Society Publishers is EU Compliant.
See newsociety.com for more information.

LIBRARY AND ARCHIVES CANADA CATALOGUING IN PUBLICATION
Title: Fruiting vegetables : a grower's guide / Jean-Martin Fortier; translated by Laurie Bennett; edited by Pierre Nessmann with the collaboration of Mathilde Lebecq and Damien Terral; illustrations by Flore Avram.
Other titles: Légumes fruits. English
Names: Fortier, Jean-Martin, author | Bennett, Laurie, translator. | Nessmann, Pierre, editor | Lebecq, Mathilde, editor. | Terral, Damien, editor. | Avram, Flore, illustrator.
Description: Series statement: Grower's guides from the market gardener; 5 | Translation of: Les légumes fruits.
Identifiers: Canadiana (print) 2025031603X | Canadiana (ebook) 20250316048 | ISBN 9781774060193 (softcover) | ISBN 9781550928129 (PDF) | ISBN 9781771424080 (EPUB)
Subjects: LCSH: Vegetable gardening. | LCSH: Vegetables.
Classification: LCC SB320.9 .F6713 2026 | DDC 635—dc23

Funded by the Government of Canada | Financé par le gouvernement du Canada | Canada

New Society Publishers' mission is to publish books that contribute in fundamental ways to building an ecologically sustainable and just society, and to do so with the least possible impact on the environment, in a manner that models this vision.

Praise for the Grower's Guides from the Market Gardener series

Jean-Martin Fortier has done it again! Incredibly beautiful, this series of pragmatic and professional instructional manuals makes it easy for anyone to start market gardening at a professional and regenerative level with a tailored focus on each aspect.

— Matt Powers, author, *Regenerative Soil* and *Regenerative Soil Microscopy*

Applying a market gardening mindset to your home garden will improve your yields and greatly reduce the amount of work involved. To do this, look no further than this series from Jean-Martin Fortier who is a pioneer in regenerative, biointensive, and organic market gardening.

— Rob and Michelle Avis, Verge Permaculture / 5th World, co-authors, *Building Your Permaculture Property*

Praise for the Work of Jean-Martin Fortier

This is a thorough farming manual that lays out a human-scale farming system centered on good growing practices and appropriate technology. Had I read this book when I was a starting farmer, I would now be farming with a walking tractor on an acre and hailing Jean-Martin as my market gardening guru! This book is going to inspire new farmers to stay small and farm profitably.

—Dan Brisebois, farmer, Tourne-Sol Cooperative Farm, author, *The Seed Farmer*

This is a fantastic addition to any aspiring market gardener's library, and even has a few new ideas for old hands. Jean-Martin has laid out all of the basics for how we can farm more profitability, productively, and passionately on a more human-sized scale.

—Josh Volk, Slow Hand Farm, Portland, Oregon

Jean-Martin Fortier extols the virtues of being small-scale, and expertly details the use of such scale-appropriate tools as broadforks, seeders, hoes, flame weeders, low tunnels, high tunnels, and many other unique tools, specifically designed for this brand of farming. [He provides] beginning farmers a solid framework of the information they need to start up and become successful small-scale organic growers themselves.

—Adam Lemieux, Product Manager of Tools & Supplies, Johnny's Selected Seeds

Creating a future where humans live in harmony with nature and with each other

Founded by Jean-Martin Fortier, the Market Gardener Institute is committed to inspiring and supporting new organic growers at every stage of their journey. Our mission is to equip them with the essential technical skills needed to thrive in their vital agricultural work.

Our vision is to multiply the number of organic, regenerative farms around the world and create a future where humans live in harmony with nature and each other.

www.themarketgardener.com

Presenting the collection

Grower's Guides from the Market Gardener

Hi!

I am delighted to bring you this new collection of practical guides. The advice you'll find in these books is based on working methods I developed on my own microfarm and refined over the last two decades. While plenty of these concepts are not new and were passed on to me by different mentors through the years, many other ideas stem from my own farming experience. I am sure you'll come across a number of tips and tricks that are innovative, proven, and easy to implement.

Whether you are a home gardener, hobby farmer, new market gardener, or an experienced farmer looking to transition to more intensive growing on smaller plots, you will find everything you need to take your horticultural practices even further.

Wishing you success and happiness in your agricultural adventures!

Jean-Martin Fortier, market gardener in Saint-Armand, Quebec

Contents

Introduction: A Few Words About My Background

Drawing on principles from agroecology, permaculture, and entrepreneurship, I champion a modern form of nonmechanized farming, carried out on a human scale.

On a human scale means feeding many local families, while respecting the human and natural ecosystems in which we operate.

On a human scale means allowing market gardeners to make a decent living from their work, to run their businesses as they see fit, and to give themselves more time off than conventional farmers.

On a human scale means evolving through the use of technology but especially by relying on people and their skills and knowledge.

From Organic Farms...

I studied agroecology at McGill University's School of Environment in Montréal, where I met my wife and business partner, Maude-Hélène Desroches. At the time, we were both looking to create a new model for farming, one that would have a positive environmental impact. After graduation, we spent two years in New Mexico, USA, working on an organic farm and learning to be market gardeners.

Our microfarming aspirations were later fueled by a trip to Cuba where we spent time on *organopónicos*, fascinating urban farms that were established during the American embargo. During that era, after the fall of the USSR, the country developed a biointensive and urban agricultural model to ensure food security for the island's residents.

... to a Family-Run Microfarm

Back in Quebec in 2004, we acquired a small plot of 10 acres in Saint-Armand, in the scenic Eastern Townships. On this land, we experimented with our innovative approach to market gardening, which especially drew from the work of Eliot Coleman, an American market gardener who has been highly influential in the world of organic microfarming.

We built a 2-acre market garden, Les Jardins de la Grelinette, where we were able to test the first iterations of my method, now called the Market Gardener Method. It consists of crop rotation, the near-exclusive use of hand tools, organic growing practices, and shorter marketing channels, with direct sales made through CSA boxes and farmers' markets. At Les Jardins de la Grelinette, Maude-Hélène and I both worked full-time, and hired two farm workers (one full-time and the other part-time) to help with harvests.

Making 2 Acres Profitable

Success came quickly, both in terms of harvests and direct sales. After bringing in $33,000 in our first year, we earned twice that in the following year, and more than $110,000 in our third year of operation.

We were thus able to earn a living as market gardeners from almost the very beginning. Since then, our farm has continued to feed more than 200 families every year, offering roughly 40 types of vegetables, all grown on just 2 acres. Over the years, our harvests expanded and sales continued to increase. Eight years after starting the farm, I presented this farming model in a practical guide called *The Market Gardener* in 2014. The book was an instant success — over 250,000 copies have now been sold, and it has been translated into nine languages.

In 2015, with the support of a generous patron, I founded Ferme des Quatre-Temps in Hemmingford, Quebec, with the vision of creating a model for the future of ecological agriculture. On this 160-acre farm, we established a polyculture system in a closed-loop cycle, raising pasture-fed cattle, pigs, and hens, alongside a culinary laboratory. At the heart of the farm, 7.5 acres were dedicated to a market garden, where we applied the growing methods developed at Les Jardins de la Grelinette. It is here that I teach my apprentices the principles of productive and profitable market gardening.

The project was featured in a TV show called *Les fermiers*, which follows the evolution of Ferme des Quatre-Temps and its apprentices, who later start their own farms in front of the cameras. The show was a hit in Quebec and is now available on TV5 Monde and Apple TV.

In parallel, I worked to expand my methods to reach a broader, global audience. In 2018, we launched the Market Gardener Masterclass, a fully online course now available in over 90 countries. To further support this initiative, I founded the Market Gardener Institute with a clear mission: to educate the next generation of growers by equipping them with the knowledge, skills, and resources needed to become leaders in the organic farming movement.

The Institute has two key objectives: to teach best practices in market gardening techniques and growing methods, and to demonstrate that small-scale farming worldwide can not only be ecological but also productive and profitable. On a global scale, it's the number of farms, not their size, that holds the key to feeding the world.

Inspiring Change

My ambition is to drive meaningful change in society by promoting a way of farming that honors nature, supports communities, and empowers local farmers. I believe in a decentralized farming model, built farm by farm, as the foundation for a truly sustainable and resilient food system.

Since 2020, I have proudly served as an ambassador for the prestigious Rodale Institute, which researches regenerative organic farming practices in the United States and beyond. I am also honored to be the ambassador for Growers and Co., a company that develops tools and apparel for new organic growers. In 2023, we launched Espace Old Mill, a restaurant and market garden set in one location. The restaurant uses the best produce in the region, including harvests from our own farm.

What Is the Market Gardener Method?

While my approach may seem innovative, it is founded on practices that were first developed by 19th-century Parisian gardeners, who fed more than two million people through a network of thousands of market gardens—precursors to our modern-day microfarms—within the city of Paris.

These market gardeners applied remarkable ingenuity, skills, and knowledge to meet the increasing food demands of a city in the midst of urbanization and demographic expansion. They achieved this through organic, nonmechanized agriculture. From the mid-18th century to the 20th century, many books were written about the innovative practices of these market gardeners, whose technical feats were admired throughout Europe. But with the advent of modern practices, much of this know-how was relegated to the past.

As a result of mechanization, the advent of agronomic science, and improved refrigeration and transport that brought in fresh and inexpensive food grown abroad, farms grew in size, became less diversified, and took on a more technological focus—a trend that continues today.

Fortunately, these inspiring models led to the development of horticultural methods that have endured, and with the same objective: to grow sustainably, by maximizing vegetable yields without degrading soil quality. We now use the term "biointensive" to describe these methods. Unlike extensive agricultural operations, they continue to work on a human scale and offer farmers the opportunity to use little mechanization. Despite what some may believe, this approach is also profitable.

By working on only small plots of land, market gardeners can keep start-up investments to a minimum, compared to the funds needed for a conventional farm. Biointensive farmers also require a smaller workforce, doing the work themselves with the help of just a few employees. They also sell their produce directly to customers, avoiding commissions to intermediaries. These three factors allow market gardeners to start generating profits quickly.

Still, it's important to remember that working the land is never easy. While market gardeners can make a good living with this method, the first seasons are time-consuming and require a significant workload and financial investment. In this profession, nothing comes easy, and every dollar you earn is the fruit of your labor, the result of your organizational skills. That's why I always tell my apprentices to learn how to work smarter, not harder.

From a financial perspective, market gardeners should plan to start with an investment of $50,000 to $150,000, depending on whether certain assets are already available—such as a building that can be converted, access to abundant water, electricity, natural gas, or a vehicle. This amount does not include the cost of purchasing land, which can be amortized over 20 years, if needed. Renting is also an option that can prove very profitable, especially when the farm is located near a city or an affluent municipality, where land is expensive.

Regardless of experience and preparation, the first years of market gardening will be intense. Opening new ground, constructing greenhouses and tunnels, and setting up infrastructure (irrigation, washing and packing stations, nurseries, etc.) all take extra time and effort. However, once this phase is complete, market gardeners who have mastered their craft can do more than just make a living off a few acres—they can earn a very decent living.

This leads to another key principle I teach: your farm should work for you, not the other way around. Profitable and productive farming is possible, but you need to set it up for success.

Preface: Unearthing a World of Fruiting Vegetables

July is a time of plenty in every veggie garden! In this summer month, plant growth peaks, and the garden is lush and thriving. When all our beloved vegetable crops are basking in the sun's warmth, it's time to harvest them: zucchinis, cucumbers, beans, peas, and of course tomatoes, peppers, and my personal favorites, watermelons and melons. Their vibrant colors evoke the warm summer nights and lovely days spent gardening in the sunshine. What a joy!

However, as any market gardener will tell you, fruiting vegetables are some of the most demanding crops. You must learn to prevent pest damage, anticipate numerous diseases, such as late blight, and manage the irrigation of each crop, providing enough water but never too much. Their sowing and transplanting stages also require a specific expertise and skill set. The vagaries of weather and fertilization needs are additional challenges to be tackled properly.

In this guide, I'm sharing my best production practices, tips, and techniques that rely heavily on a preventive approach. The key is understanding the needs of each vegetable and taking a methodical approach to crop maintenance.

May this journey, deep into the world of fruiting vegetables, be a source of inspiration for your next gardening season. Together we can grow these vegetables successfully, even under the hottest sun, contributing to a better future for everyone. Good luck!

Jean-Martin Fortier, market gardener in Saint-Armand, Quebec

Fruiting Vegetables: The Essentials

Fruiting vegetables and seed vegetables make up a highly diverse plant category that significantly contributes to our diet. It encompasses species from winter squash to beans and the Solanaceae family, including tomatoes—the quintessential crop—and offers unique flavors and varied textures, making these vegetables true culinary gems.

With fruiting or fruit vegetables, the part that we eat is in fact the mature fruit of the plant, which is a swelling of a fertilized ovary. Once pollinated, the flower begins forming a fruit containing the seeds needed for plant reproduction. This complex botanical process generates myriad vegetable shapes, sizes, and textures. Think of tomatoes, eggplants, sweet peppers, hot peppers, tomatillos, and ground cherries. In contrast, winter and summer squash, cucumbers, and melons produce what is called false or accessory fruit, which forms when several flowers fuse into a single fleshy organ.

Legumes, or Fabaceae, such as beans and peas deserve a prominent place in our diets because they deliver many nutritional benefits, store well, and benefit the soil. These vegetables are eaten in two ways, either unripe (pods and seeds), such as green beans, green peas, snap peas, and snow peas, or ripe and dried, as with flageolet beans, dried peas, and broad beans.

Nutritionally legumes are a rich source of plant protein, fiber, vitamins, and essential minerals. When regularly incorporated into a diet, they can contribute to maintaining a healthy nutritional balance, providing nutrients that support muscle health, digestion, and general vitality.

Once ripe, the dry seeds of beans and peas can also be harvested and stored for later consumption, especially during the winter. This provides a significant advantage, allowing gardeners to stock their pantry with protein stores to cook over the coming seasons.

Winter and summer squash are iconic fruiting vegetables with succulent flesh and an array of shapes, colors, and flavors: the delicate patty pan, the mild red kuri, the creamy butternut, or the less common chilacayote squash. When stored properly, winter squash can keep almost until the following year's first harvests, providing valuable nutrients in the offseason.

With diverse origins dating back to ancient times, fruiting vegetables and seed vegetables represent human migrations, cultural interactions, and botanical crossbreeding. All types of squash, from patty pan, red kuri, and butternut to chilacayote to zucchini (courgette), hail from South and Central America, and were later introduced to Europe. Cucumbers are native to Asia, where these vegetables were grown for millennia before conquering kitchens the world over.

Melons have a long history in Asia, Africa, and the Middle East and only recently became staples in European and North American markets. Beans and peas were cultivated in various parts of the world, from South America to Asia, providing important sources of plant protein to different populations.

Tomatoes, native to South America, conquered Europe in the 16th century, while eggplant, sweet peppers, and hot peppers originate from tropical regions in Asia and the Americas. Physalis, a plant group that includes ground cherries, alkekengi (or "love in a cage," in French), Cape gooseberries, and tomatillos, has long been grown in South America and Africa before spreading around the world.

Beans and peas, from the Fabaceae family, have two distinguishing features: edible seed pods and, most notable, the ability to fix atmospheric nitrogen in the soil.

These plants use root nodules to form a symbiotic relationship with Rhizobium bacteria, a unique collaboration that converts atmospheric nitrogen into a form that is available to plants. In other words, beans, peas, and other vegetables in this family actively contribute to enhancing soil nitrogen, an essential nutrient for plant growth, and by extension for crops that are adjacent or will succeed them in the same bed.

When harvesting Fabaceae, you may opt for a practice that is greener and better for your soil: don't pull out the roots and don't remove and throw away all the foliage (stems, leaves, etc.). Instead, it is best to cut these aboveground parts at the soil level, leaving the root ball in place. Covered in a thriving community of bacterial nodules, the roots will continue to release nitrogen into the soil as they decompose. This method improves soil fertility and reduces the need for additional nitrogen inputs.

Fabaceae also generate a significant amount of green waste from their aboveground components. Rather than throwing this out, you can finely chop it to make a beneficial mulch that can be left in situ as an amendment or applied to the surface around other crops. It acts as a protective cover, maintaining soil moisture, regulating soil temperatures, and limiting weed growth. By using this green vegetable waste, growers turn it into a precious resource that contributes to improving overall soil health in the garden.

Soil nitrogen can therefore be significantly enhanced naturally by Fabaceae crops. To get the most out of this nutrient boost, consider planting nitrogen-loving vegetables such as leafy greens (spinach, lettuce, etc.) after Fabaceae in your crop rotation. These will benefit from the available nitrogen in the soil, which promotes vigorous growth and high yields.

Tomatoes, the undisputed VIPs in every vegetable garden, grow particularly well when given special attention. In this book series, *Tomatoes: A Grower's Guide* is dedicated to describing their stages of cultivation. Because they are immensely popular and have a rich varietal diversity, specific growing methods and techniques have been developed to get the most out of this fruiting vegetable. The guide presents detailed information about growing techniques from sowing, transplanting, pruning, and trellising to fertilizing, watering, pest management, and harvesting.

In the kitchen, fruiting vegetables are the foundation of so many classic dishes, such as ratatouille and mixed salads, as well as more unusual meals such as tian and moussaka. Provençal tian, made from thinly sliced zucchini, eggplant, and tomatoes baked with herbs, is an ode to the simplicity and freshness of Mediterranean cuisine, as is Greek moussaka, featuring layers of grilled eggplant, tomato, and ground meat.

To enjoy the advantages of these fruiting vegetables during fall and winter, canning is a practical solution. For instance, tomato sauces and ratatouille—a gourmet blend of zucchinis, eggplants, peppers, and tomatoes—are well suited to this preserving method.

Squash, cucumbers, beans, and peas are annual crops, completing their life cycle in one year. It's therefore important to sow or plant these vegetables at the right time, depending on the climate, so they will grow during the summer months.

While tomatoes, eggplants, and peppers are referred to as annuals in many regions, they can behave like perennials in milder climates. Where winters are warmer and frosts are uncommon, these Solanaceae crops can live for several years in a row. Still they may have a limited lifespan, even in mild climates, due to factors such as a decline in general vigor or the spread of disease.

After harvesting tomatoes, peppers, and eggplants, consider planting quick crops at the end of the season. Opt for fast-growing vegetables such as lettuce, baby greens, pac choi, and radishes. Because these Solanaceae crops are heavy feeders, introduce cover crops, such as phacelia or a seed blend, to your soil.

Fruiting vegetables—ambassadors of the sun—are refreshing summer delights with vibrant flavors. The term “fruiting vegetable” encompasses emblematic crops such as juicy tomatoes, sun-drenched peppers, velvety eggplants, delicate zucchinis, invigorating cucumbers, and Fabaceae as well as all types of squash, which offer diverse flavors and textures. The varieties presented in this guide, beyond the classics we all know and love, include lesser-known treasures such as Cape gooseberries, sticky nightshades, pepino melons, asparagus peas, and tomatillos. In this diversity lies a culinary richness, an infinite palette of textures and flavors for food lovers hoping to create novel dishes.

For this book, I’ve selected twenty fruiting vegetables you can incorporate into your home garden or professional garden plan.

Jean-Martin Fortier's Favorite Fruiting Vegetables

Cucurbitaceae (Cucurbits)

Most types of squash are native to Central and South America. They were brought to Europe in the 16th century and quickly gained in popularity. The genus *Cucurbita* includes about 15 species, 3 of which are found in our vegetable gardens: *Cucurbita pepo, Cucurbita moschata,* and *Cucurbita maxima*. Because they easily form hybrids, there are hundreds of varieties with different shapes, sizes, colors, and flavors. These creeping annuals and be recognized by tendrils growing along their stems. Flowers are large and yellow. Winter squash varieties have a slightly tougher skin, a range of flavors (sweet, nutty, etc.), and a longer shelf life. Summer squash includes zucchinis and patty pans, which have tender flesh and skin when harvested young. Squash are nutritious, packed with vitamins, trace elements, and minerals, as well as fiber, which aids digestion. They can be eaten in numerous ways: in soups, stuffed, mashed, baked in gratins, sauteed, fried, and more.

Winter Squash

COMMON NAMES: Squash, winter squash, gourd.
SCIENTIFIC NAME: *Cucurbita.*
FAMILY: Cucurbitaceae.
REQUIREMENTS: Winter squash cannot withstand cold or frost, so we recommend sowing it in a warm room or under cover. Crops can be grown outdoors, but should be planted after the last frosts. We highly recommend using a woven ground cover or other mulch. Choose a sunny location and loose, humus-rich soil.
SPACING: 1 row down the middle of the bed (30 in. or 75 cm wide), 24 to 36 inches (60 to 90 cm) apart.
SEEDING: In April, in a bright and heated room, like a sunroom, or on heat mats in professional operations.
TRANSPLANTING: In May as soon as possible after the last frosts.
DAYS TO MATURITY: 80 days, up to 120–150 days.
ENEMIES: Powdery mildew, anthracnose, black aphids, root aphids, slugs and snails, rodents.
VARIETIES: See the following pages.
A NOTE FROM JEAN-MARTIN FORTIER: Winter squash is typically sold over the winter and is popular with customers. It's also a low-maintenance crop. However, compared to other vegetables, it isn't the most profitable option for professional growers; winter squash is slow-growing and generates only a modest revenue, given its duration in the field. Therefore spend as little time as possible on its maintenance. You also need to harvest each variety at the right time.

Patty pan squash

Cucurbita pepo: Spaghetti, Acorn, Patty Pan, and others

DESCRIPTION: This most common species in vegetable gardens includes summer squash (zucchini or courgette, patty pan, see pages 30 to 35) and winter squash (pumpkins, spaghetti squash, etc.). You can identify them by their stem, with its sharp curve just above the fruit, and stiff, heart-shaped leaves. This type of squash is harvested and eaten before it reaches maturity.

VARIETIES: Vegetable Spaghetti Squash, Sucrière, Ebony Acorn, Baby Boo, Jack Be Little, Delicata, Melonnette Jaspée de Vendée, Pomme d'Or, Courge de Nice à Fruits Longs, Pâtisson Blanc, Patidou (or Sweet Dumpling), Thelma Sanders Sweet Potato, Table Queen.

Red kuri squash

Cucurbita maxima: Pumpkin, Red Kuri, Buttercup, and others

DESCRIPTION: *Cucurbita maxima* produces large pumpkin-like fruits. This most cold-hardy species is ideal for gardens in northern regions and has characteristics similar to *Cucurbita pepo*. The most marked difference is that *Cucurbita maxima* have cylindrical spongy stems that flare out just above the fruit. They require less heat than *Cucurbita pepo* plants and can be cultivated in cooler climates. As soon as the fruits form, they develop their final color.

VARIETIES: Potiron d'Alençon, Rouge Vif d'Étampes, Jaune Gros de Paris, Red Kuri, Buttercup, Lady Godiva, Pink Jumbo Banana, Galeux d'Eysines, Marina di Chioggia, Doux Vert d'Hokkaido (or Potimarron).

Butternut

Cucurbita moschata: Butternut, Musquée de Provence, and others

DESCRIPTION: *Cucurbita moschata* have long stems with tendrils, downy soft leaves, and short flower sepals. Fruits are often elongated and club-shaped, their color depending on the variety. With this species, yields can fluctuate significantly from year to year, while *Cucurbita maxima* and *Cucurbita pepo* are more consistent.

VARIETIES: Waltham Butternut, Black Futsu, Musquée de Provence, Tromba d'Albenga, Sucrine du Berry, Honeynut.

Chilacayote

Uncommon Squash

DESCRIPTION: *Cucurbita ficifolia* is known by many names, including chilacayote. A vigorous runner, this climbing species can reach 50 feet (15 m) long. Several rare varieties exist but are uncommon in more northern climates. Chilacayote, an ancient Mexican variety, produces oval green fruits with white spots, weighing between 4 and 13 pounds (2–6 kg). It has a mild flavor and will keep for 1 to 2 years.

VARIETY: Courge de Siam.

Growing Winter Squash

"Winter squash are popular and pair well with root vegetables in a winter meal. All you need to do is... maximize your space!"

Jean-Martin Fortier

Planting

Seeding Under Cover

Home gardeners can seed winter squash into pots in a mini greenhouse or a bright heated room. Sow the crop a month before planting, dropping three seeds per hole.

Market gardeners use plug flats to seed winter squash in a greenhouse two weeks before the last frost date, to plant seedlings after that. Place seeds about 0.5 inches (1 cm) deep in moist potting mix. Cover with a thin layer of the mix and water the surface with a gentle spray.

Plug flats can be moved to heating mats kept at 86°F (30°C) until germination. After the seeds sprout, maintain temperatures of roughly 75°F (24°C) during the day and 64°F (18°C) at night, making sure the soil is always moist.

Harden off seedlings 4 to 7 days before planting by exposing them to outside temperatures and reducing watering.

Tip from Jean-Martin Fortier

As these crops are slow-growing, plant seedings in cold climates as early as possible once the risk of frost has passed. Ideally nighttime temperatures must be at least 50°F (10°C). The fruits should therefore be mature or nearly mature when the foliage dies off in the fall.

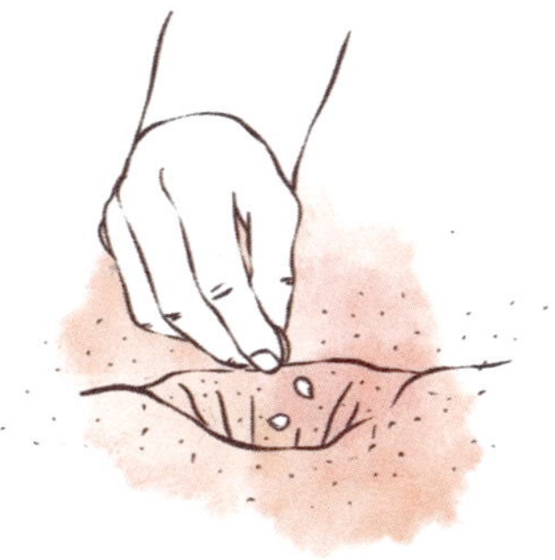

Direct Seeding in the Field

Home gardeners can sow squash seeds directly into garden soil at the start of May, dropping 3 or 4 seeds per hole. This produces more vigorous plants without the stress of transplanting. Allow 6 feet (2 m) of space in all directions for creeping varieties and 2.5 feet (80 cm) for more compact varieties. After 15 days, remove the weaker squash plants.

Preparing the Soil

After 2 to 4 weeks, once seedlings have 2 or 3 leaves, they can be planted outdoors. Loosen the soil with a broadfork and add a 2-inch (5 cm) layer of compost. Run a power harrow, set to a depth of 2¼ inches (6 cm), down the bed to break up and level the soil surface.

2

Cover beds with a woven ground cover, with premade holes, 4 inches (10 cm) in diameter, every 3 feet (90 cm). Secure the ground cover and keep it taut using fabric staples every 5 feet (1.5 m). Otherwise, spread a layer of straw, 3 to 4 inches (8–10 cm) thick, over the bed.

Tip from Jean-Martin Fortier

Squash plants require a lot of nutrients. When preparing beds, spread plenty of fresh or mature compost. Professional growers can apply a 2- to 3-inch (5–8 cm) layer down the middle of the bed, while home gardeners can drop compost into the holes before seeding. If compost was added to previous crops in the bed, apply only alfalfa meal and pelleted chicken manure.

Transplanting

Winter squash seedlings are sensitive to wind and sun, so transplant them towards the end of a cloudy day with little or no wind.

1. **Before planting, moisten the seedlings and soil. Gently pull each root ball from its pot and plant using a dibber, ensuring it's in the center of the hole. Squash seedlings are fragile, so be careful not to bury the stem; the top of the root ball should be level with the surface.**

2. **For about one week after planting, water regularly to keep the soil moist. Cover the crop with insect netting supported by metal hoops.**

Tip from Jean-Martin Fortier

For professional market gardeners, winter squash is not the most lucrative crop. To minimize your time required in the field, use a woven ground cover or other mulch to suppress weed growth and the likelihood of disease.

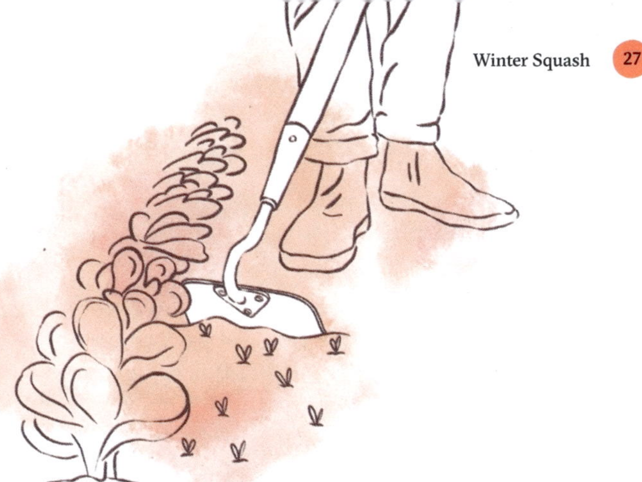

Maintenance

Irrigation
Set up the irrigation system, laying two drip lines under the woven ground cover or other mulch. This keeps the leaves from getting wet which makes the plants less prone to disease. In general, water when the soil is dry, wetting 1 inch or so (3–4 cm) down.

Weeding
If you opt to cover the beds with a straw mulch, make sure to hoe the soil surface when you first plant the crop then apply the mulch in July.

If you set up a woven ground cover, weed by hand so creeping vines don't take over the plot.

Plant Protection
Installing insect netting or a floating row cover in early spring protects seedlings from insects (whiteflies, aphids, etc.), strong temperature fluctuations, and wind. Remember to remove the netting when the first blooms appear. Squash plants are susceptible to fungal diseases, such as powdery mildew and late blight. Several practices can help prevent these diseases: crop rotation, planting in well-drained soil, removing infected plants, mulching, appropriate spacing to promote good airflow, and avoiding overhead sprinklers to avoid wetting the leaves.

As a preventive measure, you can also spray the crop with a horsetail tea solution (decoction).

Tip from Jean-Martin Fortier
Home gardeners can lay a tile under each fruit to keep it off the damp soil, preventing the bottom from rotting.

Harvest

With winter squash, harvesting is a crucial step, so make sure to follow these rules:

- Harvest each variety twice. The fruits won't all be ready at the same time, so plan harvest dates a few weeks apart.
- Harvest squash in dry weather so that the skin won't be wet.
- Wipe any dirt off the squash.
- Use pruning shears to cut the stem ¾ inch (2 cm) from the fruit.
- Don't let the crop freeze as this may result in a diminished shelf life.
- Gently place the fruits in harvest bins or wooden crates.

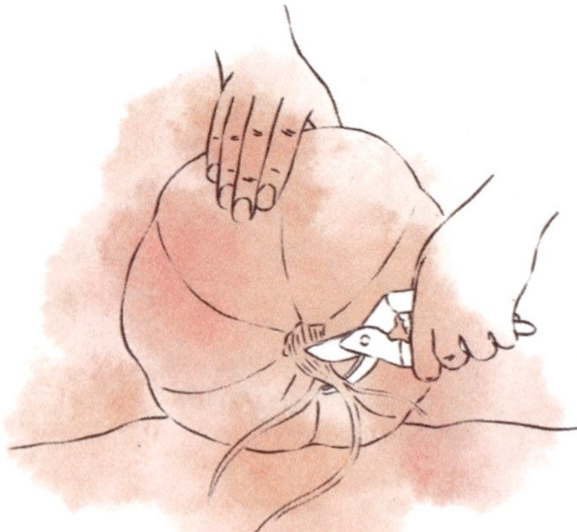

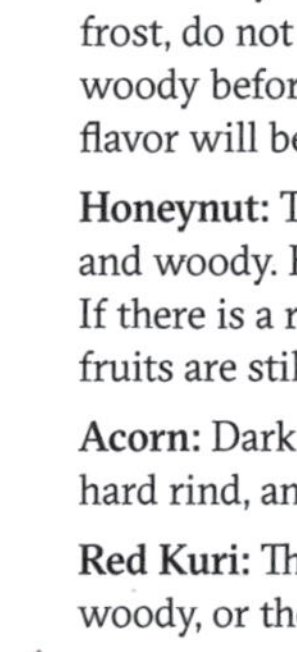

Signs that the squash has reached maturity, examples:

Musquée de Provence: Dark orange spot where the squash touches the ground.

Delicata: Dark orange line or spot where the squash touches the ground.

Butternut: The stem near the fruit is dry and woody. If there is a significant risk of frost, do not wait for the stem to turn woody before harvesting; however, some flavor will be lost.

Honeynut: The stem near the fruit is dry and woody. Fruits should mostly be orange. If there is a risk of frost, harvest when the fruits are still half-green.

Acorn: Dark green with some orange areas, hard rind, and dry woody stem.

Red Kuri: The stem near the fruit is dry and woody, or the skin is dark orange.

Spaghetti: Golden-yellow and dry woody stem.

Musquée de Provence, Red Kuri, and Delicata varieties will be ready first.

Tip from Jean-Martin Fortier

When harvesting winter squash, handle each fruit as if it were an egg as they are quite delicate. When bruised, even lightly, they begin to rot.

Storage

1 Store squash for 2 to 3 weeks in a closed space maintained between 80°F and 90°F (27–32°C). Professional growers can leave them in a greenhouse or nursery while home gardeners can use any enclosed space kept at the right temperature. Be careful to avoid temperature variations to prevent condensation on the fruits. Delicata and Musquée de Provence varieties do not need to be cured.

2 Store cured squash for the winter on shelves or in storage bins in a dry space kept between 50°F and 60°F (10–15°C). Use a dehumidifier to maintain humidity at 50%.

Home gardeners can store squash in a frost-free cellar on a bed of straw all winter long.

Tip from Jean-Martin Fortier

Make sure to ventilate your storage space to promote good air circulation. Stored properly, winter squash can keep for more than 2 months! Each variety has a different shelf life.

Summer Squash

In the Cucurbitaceae family, *Cucurbita pepo* is the most well-known and widely consumed, thanks to zucchini (courgette), which is prolific and one of the easiest crops to grow. Squash are native to Central America and first landed in Europe early in the 16th century. In the 18th century, cultivation took a new turn when Italians began harvesting the crop early, resulting in the zucchini we know today. This annual herbaceous plant with vines that bear fruits in various shapes, colors, and sizes can be recognized by its raspy, rigid stems. Flowers are yellow. Summer squash contain vitamins A and C, improve digestion, have nutritional and diuretic properties, and provide calcium, phosphorus, and iron. The low-calorie fruits tend to be slightly sweet and can be eaten in many ways: sautéed, stuffed, au gratin, in soup and ratatouille, etc.

Summer Squash

COMMON NAMES: Zucchini, courgette, patty pan.
SCIENTIFIC NAME: *Cucurbita pepo.*
FAMILY: Cucurbitaceae.
REQUIREMENTS: Zucchinis thrive in soil that is rich, cool, light, and well-drained in warm, temperate climates with full sun.
SPACING: 1 row down the middle of the bed, with seedlings set 24 inches (60 cm) apart.
SEEDING: Sow in April if under cover, in a mini greenhouse, sunroom, or warm and bright room, or on heating mats (for professional growers). Direct seed in the field from early May to July.
DAYS TO MATURITY: 60 to 90 days.
ENEMIES: Powdery mildew, anthracnose, black aphids, root aphids, slugs and snails, rodents.
VARIETIES: Verte Non Coureuse des Maraîchers, Coucourzelle, Gold Rush, Genovese, Ronde de Nice, Grisette de Provence.
A NOTE FROM JEAN-MARTIN FORTIER: Zucchinis are easy to grow, productive, and long-lasting. Stagger sowings to ensure a consistent harvest throughout the season.

Growing Zucchinis

"It's the variety among summer squash cultivars that makes this crop so attractive and popular at the market and in home gardens."

Jean-Martin Fortier

Planting

Seeding Under Cover

Home gardeners can sow zucchinis in a mini greenhouse or a heated bright room. Seedlings can be grown in pots (sow 3 seeds per hole) and later planted after the last frosts. Keep only one of the strongest of the three seedlings.

Seeding in Greenhouses

Professionals should sow zucchinis in a greenhouse 15 days before the transplant date. Sow into plug flats filled with lightly moistened potting mix. Cover with a thin layer of mix and water with a gentle spray. Place seed flats on heating mats to keep the soil temperature at 86°F (30°C) until germination. Once the seeds have sprouted, maintain temperatures of roughly 75°F (24°C) during the day and 64°F (18°C) at night, keeping the soil moist at all times.

About 7 days before planting, place seedlings outdoors under an awning or floating row cover to harden them off.

Preparing the Soil

Zucchini plants grow well in rich soil. Spread an even layer about 2 inches (5–6 cm) deep of compost. Add a nitrogen-rich fertilizer (mixture of alfalfa meal and pelleted chicken manure) that will quickly help the seedlings establish roots.

Planting

Clear the surface with a bed preparation rake, loosen the soil with a broadfork, and amend with compost and fertilizer. Run a power harrow down the bed at a depth of 2 inches (5 cm).

2

Dig a hole with a dibber, remove the seedlings from the pots, and plant the root balls. Zucchini seedlings are quite fragile, so handle them carefully and make sure that the top of the root ball is level with the soil surface. Do not bury the stem.

Give each plant a good watering to settle the soil around the root ball. Set up metal hoops and lay insect netting over the crop. Once the first flowers appear, remove the netting to allow for pollination.

Maintenance

Weeding

If you are not using a woven ground cover, you must hoe the beds about every 15 days to loosen the soil surface and eliminate weeds. At the end of June, lay down 3 to 4 inches (8–10 cm) of plant mulch.

Irrigation

Frequent watering is essential for consistent fruit development and uniform size. When plants are flowering water thoroughly without wetting the foliage. Market gardeners should lay four lines of drip tape down the bed, securing them with fabric staples. Home gardeners can use a hose or watering can to direct water towards the base of each plant.

Plant Protection

To protect seedlings from insects (whiteflies, aphids, etc.) and limit the drying effect of wind, lay insect netting or a floating row cover over metal hoops in the beds.

Zucchinis are sensitive to fungal diseases, such as powdery mildew and late blight. To keep disease at bay, establish a crop rotation, plant in well-drained soils, remove infected plants, and apply a mulch. Maintain proper spacing between seedlings to promote good airflow and avoid using a sprinkler to keep leaves from getting wet. A horsetail solution (decoction) can be used as a preventive treatment.

Harvest

Picking

When fruits are 6 to 8 inches (15–20 cm) long, harvest them using a sharp knife (Opinel type) to cut the stem just under an inch (1–2 cm) from the fruit. Wear long sleeves to avoid contact with the prickly hairs along the stem that can cause rashes. When harvesting blossoms, store them in the fridge immediately to prevent wilting. Wipe any dirt splatters off the zucchinis with a cloth.

Storage

Store zucchinis in crates in a cool room maintained at 54°F (12°C) or in your refrigerator's vegetable drawer. Market gardeners should store the crop for no more than 4 days before selling them.

Cucumbers

Cucumbers originated in India then quickly spread throughout Asia and Africa in antiquity. Their arrival in France was recorded as early as the 9th century when Charlemagne, and later Louis XIV, decreed that it should be cultivated. Cucumbers became popular in the 20th century as a result of greenhouse cultivation. This annual herbaceous creeping vine can reach 8 feet (2.5 m) long, with alternate leaves and tendrils on stems that are branched, angular, and voluble. Depending on the variety, fruits may be elongated, round or oval, smooth or spiny, and green, whitish, or yellowish. Rich in essential minerals like calcium, iron, magnesium, manganese, phosphorus, and zinc, cucumbers improve digestion and have diuretic, hydrating, and nutritional properties. They also contain vitamins A, B1, B2, and C. Raw cucumbers can be as crudités and in salads. They can also be cooked in gratins and soups, and stuffed, or served with a sauce. Pickled cucumber is an excellent condiment for cold cuts, charcuterie, and stews.

Cucumbers

COMMON NAMES: Cucumber, gherkin.
SCIENTIFIC NAME: *Cucumis sativus.*
FAMILY: Cucurbitaceae.
REQUIREMENTS: Cucumbers prefer rich, deep loose soils. They thrive in warm environments. We recommend growing them under cover.
SPACING: Plant Lebanese or snacking cucumber seedlings in 1 row per bed, 12 inches (30 cm) apart; 18 inches (45 cm) apart for English or long cucumbers.
SEEDING: In April, in a sunroom, on heating mats in a greenhouse, or in a cold frame.
TRANSPLANTING: In May as soon as possible after the last frosts.
DAYS TO MATURITY: 70 to 90 days.
ENEMIES: Anthracnose, gray rot or botrytis, powdery mildew, slugs, black aphids, and root aphids.
VARIETIES: Cucumbers grown under cover: Katrina F1 (small Lebanese cucumber), Kalunga F1 (English or seedless cucumber). Field cucumbers: Blanc Hâtif (the easiest to digest), Vert Long Maraîcher (excellent yields), Le Généreux (for pickling, if harvested early), Marketmore (early, productive, no bitterness). For pickling: Fin de Meaux (hardy, early, and productive), Hokus (very productive and disease resistant), Cornichon de Bourbonne (productive and vigorous), Vert Petit de Paris (very productive, flavor is quite delicate).
A NOTE FROM JEAN-MARTIN FORTIER: After tomato crops, cucumbers deliver the highest yields.

Growing Cucumbers

"Growing in tunnels with a stable and controlled climate significantly limits damage caused by disease."

Jean-Martin Fortier

Planting

Seeding into Pots

About 4 weeks before the last frosts, fill 3- to 4-inch (8–10 cm) pots with potting mix and sow 1 to 3 seeds in each, at a depth of about 0.5 inch (1cm).

2

Home gardeners can put the pots in a warm, bright room or cold frame, while professional gardeners can place them on heating mats set to 77°F (25°C) in a greenhouse. Keep the soil moist and maintain a consistent relative humidity. Water regularly with a gentle spray. After seeds sprout, maintain temperatures of roughly 64°F to 66°F (18–19°C) both day and night. Keep only the most vigorous seedling per pot.

Tip from Jean-Martin Fortier

To promote germination without a heating mat, you can cut the top off a plastic bottle to make a mini greenhouse that creates a warm and humid atmosphere. Remove it as soon as the seeds germinate. Home gardeners without a sunroom can also purchase indoor mini greenhouses, at garden centers, that provide ideal growing conditions.

Seeding into Flats

1 Spread a layer of potting mix in a seed flat, then sow thinly with a seeder. Use a sieve to cover the seeds with ⅛ inch (3 mm) of fine potting mix.

2 Lightly compact the soil with a tamper, then water with a gentle spray. Market gardeners should keep seeded flats in a greenhouse, if possible, on heating mats set to 77°F (25°C). Home gardeners can place them in a cold frame or warm room. Water regularly to keep the soil moist, but not excessively, until the seeds germinate.

Tip from Jean-Martin Fortier

Once the seedlings have germinated, allow the soil to dry out between waterings to help prevent damping-off caused by excess moisture. Watering should ideally be done in the morning or afternoon so the soil can drain before nightfall.

Potting Up

Transplant 15 days after sowing into 5- to 6-inch (12–15 cm) pots filled with potting mix. Cucumber seedlings are delicate, with a tender stem that can easily snap, so handle the root ball very carefully and make sure you don't bury the stem. For the next 15 days, store them in a greenhouse or room kept between 64°F and 66°F (18–19°C).

Direct Seeding

After the last frosts, from mid-May to early July, sow cucumbers directly into garden soil, dropping 3 or 4 seeds per hole. These plants will be more vigorous because they have been spared the stress of a transplanting. Fifteen days after direct sowing cucumbers, remove the weaker seedlings, leaving behind only the strongest and most vigorous.

Tip from Jean-Martin Fortier

Two weeks before planting cucumbers, lay a solarization tarp over the beds to warm the soil to 63°F (17°C). Transplant seedlings once outdoor temperatures are above 61°F (16°C) day and night. The average temperature in the greenhouse or tunnel should ideally be 72°F (22°C) day and night.

Transplanting

In the field, prepare the beds when cucumber seedlings have 2 or 3 leaves. Loosen the soil with a broadfork and add a layer of compost and chicken manure. Next run a power harrow down the bed, set to a depth of 8 inches (20 cm), then carefully level the surface with a rake. When growing cucumbers in a greenhouse or tunnel, tie trellis lines to the metal structure.

1 **Use a stringline or a hoe dag to mark the middle of the bed. Plant seedlings 18 inches (45 cm) apart for long cucumbers and 12 inches (30 cm) for smaller varieties and trellised crops. If you want the plants to grow along the ground, space them 40 inches (1 m) apart.**

2 **Gardeners using drip irrigation should set up two drip lines per bed, on each side of the row. Then cover the aisles and beds with a silage tarp, black side facing down.**

3 Make holes in the tarp with a knife, blow torch, or hole burning tool. Use a dibber to dig a hole the width of the pot, plant a seedling, fill it in, and water thoroughly to pack the soil around the root ball.

4 Use a tomato clip to connect the main stem of each seedling to a trellis line hanging from the overhead metal structure. Alternatively, gently wrap the string around the stem, but not too tight. Over the following weeks, pinch off suckers, tendrils, and the first fruits so the seedlings can focus energy on developing the root system and main stem.

Tip from Jean-Martin Fortier

When growing cucumbers outside, gardeners can train the plants with 5-foot (1.5 m) stakes set 6 feet (2 m) apart. Install large-mesh PVC netting across the stakes for their tendrils to latch onto. You can also build a tripod structure by driving 3 stakes into the ground (3 feet or 1 meter apart) and securing their tops with wire.

Maintenance

Irrigation

At peak production, each plant requires 3 to 4 quarts (3–4 L) per day. Make sure to water regularly, always keeping the soil moist. In the first few days, it's best to do this in the morning. Later, water during the day with a low-flow spray directed towards the ground.

Fertilization

Apply a fertilizer, made from alfalfa meal or chicken manure, along the base of the plants every 15 days. Spread it over moist soil and water immediately so the nutrients can seep into the ground.

Pruning

Home gardeners can cut the main stem above the third leaf, repeating the operation for every side shoot.

Professional gardeners should remove suckers, tendrils, fruits, and flowers until the plant is 2 feet (60 cm) tall to direct energy towards root development. Starting in the third week, remove every other fruit for long varieties. Use pruning shears to cut as close to the main stem as possible. With smaller cucumbers and pickling cukes, leave all fruits on the plant.

Trellising

Every 2 or 3 days, trellis the stems with clips or wind them around the overhead lines. Once they are about 6 feet (2 m) tall, remove all foliage from their bottom 12 inches (30 cm) to promote better airflow. Then, every week, remove 1 or 2 leaves per plant, starting from the bottom.

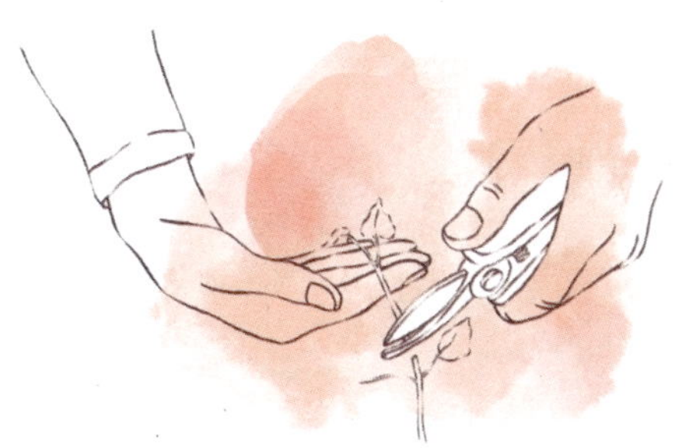

Weeding

If you did not use a silage tarp, we recommend weeding the crop every week using a stirrup hoe, which also loosens the soil and limits surface compaction. To avoid all weeding tasks, cover cucumber beds with a 4-inch (10 cm) layer of mulch in early July.

Plant Protection

One week after planting, some roots might be damaged by pythium, a fungal disease. Plan to have extra seedlings on hand to replace diseased plants, which must be removed before they contaminate the remaining crop.

To prevent powdery mildew, avoid excessive nitrogen inputs, remove affected leaves, and apply a blend of flowers of sulfur and powdered lithothamnion.

In greenhouses, you can introduce ladybugs, lacewings, and nematodes to control mites and thrips, and use pheromone traps.

Harvest

Harvest 3 times a week at the start of season and daily during peak production. Hold the ripe cucumber in one hand while using pruning shears to cut the stalk just above it. Remove the flower from the end of each fruit as it can cause rot.

Storage

If the cucumbers are not being used immediately, store them in crates in a cold room kept between 46°F and 54°F (8–12°C).

Home gardeners can store them in the refrigerator's vegetable drawer for 1 week, or when pickled in vinegar or brine, with spices, they will keep up to 6 months.

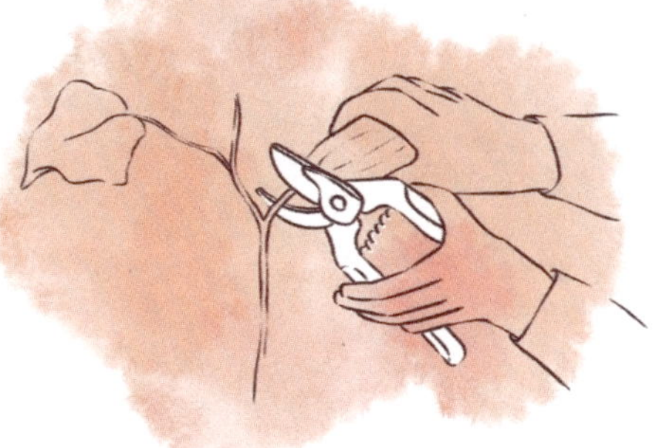

Tip from Jean-Martin Fortier

In greenhouses, misting helps keep relative humidity high (60%) and protects against spider mites, which prefer drier environments. Be careful not to mist too heavily, however, as this could foster the development of powdery mildew, which thrives in warm humid conditions.

Focus on *Cucumis anguria* and the Squirting Cucumber

Cucumis anguria, also known as the West Indian gherkin or maroon cucumber, originated in southern Africa and was later domesticated in the West Indies. Fruits are bristly but not prickly, grow to the size of a small egg, and look like a horse chestnut burr. Stems are branched and grow to be 6 feet (2 m) long. Fruits are harvested and eaten when half ripe. You can eat them like cucumbers, raw or cooked, in salads. They are sweeter and less bitter than regular cucumbers.

Seed *Cucumis anguria* into pots or flats, dropping multiple seeds per hole, around mid-April. Add compost when planting in a greenhouse or in the field. To promote good fruit development, we recommend supporting the crop with a mesh panel, stakes tied into a teepee, or lines tied to a greenhouse structure.

Squirting cucumber *(Ecballium elaterium)*, also known as exploding cucumber, belongs to the Cucurbitaceae family. It is native to the Mediterranean region, where it is a hardy perennial. In other areas, if the winter isn't overly harsh, it can stay outside. The plant has yellow hermaphroditic flowers and hairy stems that trail along the ground or, when supported, can climb up to 5 feet (1.5 m).

This crop is strictly ornamental as the fruits are poisonous. (Watch out: this variety is a botanical curiosity and not an edible vegetable! Eating it can even be fatal.) The squirting cucumber has one notable characteristic—it explodes! When the fruit reaches maturity, internal pressure makes it burst off the stem, shooting seeds several yards away. You can even make it explode by pressing on a ripe fruit with a stick.

In the spring, sow seeds in pots kept under cover in a warm and sunny location. Plant seedlings in the garden after the last frosts, ideally with good sun exposure. However, be careful not to let this plant spread its seed far and wide or it will sprawl.

Melons

First cultivated in Asia thousands of years ago, melons were later grown in ancient Egypt. In the Middle Ages, they disappeared then showed up in France towards the end of the 15th century. These annuals are grown for their sweet fruits that can be round or oval, ribbed or verrucose. They are juicy and refreshing, in colors ranging from yellow and orange to green, depending on the variety. Melons have creeping flexible stems that grow to 5 feet (1.5 m) long with large heart-shaped leaves. The fruits contain vitamins (A, B, and C) and carotene and are rich in potassium, water, calcium, copper, magnesium, phosphorus, and zinc. They have diuretic properties but are not recommended for people with diabetes. Melons are usually eaten as a snack, an appetizer, or a dessert (in a sorbet), usually raw and sliced. You can also soak them in port, Pineau des Charentes or Ricard, a delicious addition to a fruit salad. Young fruits can be preserved in vinegar or made into jam.

Melons

COMMON NAMES: Melon, muskmelon, honeydew, cantaloupe.
SCIENTIFIC NAME: *Cucumis melo.*
FAMILY: Cucurbitaceae.
REQUIREMENTS: Melons grow well in temperate-to-warm climates. We recommend planting them in caterpillar tunnels or greenhouses.
SPACING: 1 row down the middle of the bed, with plants set 30 inches (75 cm) apart.
SEEDING: In mid-April in a bright and heated room, such as a sunroom, for home gardeners. Professional market gardeners can sow crops in early to mid-May on heating mats in a greenhouse.
TRANSPLANTING: Transplant seedlings 2 to 4 weeks after sowing (mid-May below the 45th parallel or end of May in a greenhouse or low tunnel, north of the 45th parallel).
DAYS TO MATURITY: 120 to 150 days, depending on the climate.
ENEMIES: Late blight, botrytis, powdery mildew, slugs, snails, rodents.
VARIETIES: Boule d'Or (pale green flesh, keeps well), Ananas d'Amérique (10–18 oz., 300–500 g), Charentais (the most common variety in France, very sweet, flavorful orange flesh), Cavaillon (late variety, pink flesh), Hale's Best Jumbo (sweet, pink, weighing up to 3.5 lb., 1.5 kg), Petit Gris de Rennes (orange flesh, weighing 1–2 lb., 500–800 g), Cantaloupe, etc.
A NOTE FROM JEAN-MARTIN FORTIER: When growing melons, you want to produce the tastiest fruit possible. Anyone who's ever eaten a ripe garden-fresh melon will surely be back for more!

Growing Melons

"Clients love melons, although they take up a lot of space in the garden and provide only modest returns."
Jean-Martin Fortier

Planting

Seeding Under Cover

Home gardeners can sow melons in pots 1 month before the last frosts in a cold frame or bright heated room.

Professional market gardeners can sow plug flats 2 weeks before the last frosts at a depth of 0.5 inch (1 cm) and cover seeds with a thin layer of potting mix. Water with a gentle spray then keep trays in a greenhouse on heating mats to maintain a temperature of 86°F (30°C) until germination.

After the seeds sprout, maintain temperatures of roughly 75°F (24°C) during the day and 64°F (18°C) overnight, making sure the soil is always moist.

Preparing the Soil

In a greenhouse bed, rake up debris and loosen the soil with a broadfork. Apply a 2-inch (5 cm) layer of compost and spread a fertilizer such as alfalfa meal or poultry manure. Use a power harrow to level the surface. Lay a woven ground cover over the beds, secured with fabric staples.

Tip from Jean-Martin Fortier

For home gardeners who don't own a greenhouse, it is essential to install a flexible or rigid low tunnel right after transplanting the melons. If growing north of the 45th parallel, keep this in place throughout the season to ensure a good harvest. Further south, we recommend using a low tunnel only in the early stages after planting.

Planting in a Greenhouse

1. Harden off seedlings and reduce watering about 4 days before planting. Make sure both the soil and air temperatures are 70°F (21°C). At night, you should use a row cover to keep the plants from being exposed to temperatures below 60°F (15°C).

2. Carefully remove seedlings from pots or plug flats and bury them every 30 inches (75 cm) in a single row, keeping the top of the root ball level with the soil. As with other cucurbits, do not bury the stem. Water plants individually to settle the soil before setting up your drip irrigation system.

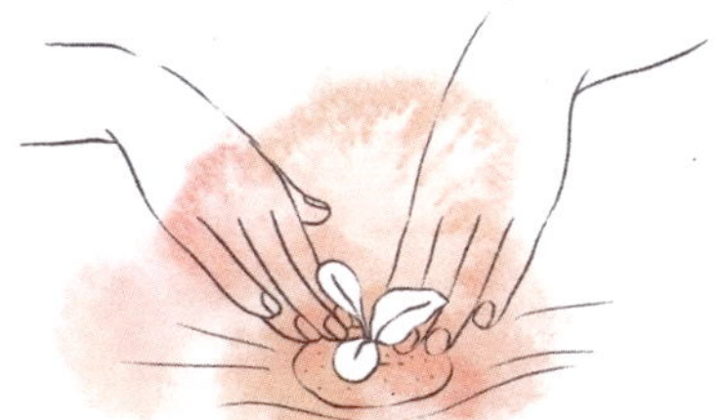

3. If you don't have a fabric mulch, plant seedlings straight into the ground. Dig a hole with a dibber, insert the root ball, fill in the hole, and lightly tamp the soil without crushing the roots. Water seedlings regularly so they never dry out.

Tip from Jean-Martin Fortier

To help the seedlings get established, a soil temperature of 70°F (21°C) is essential. You may need to lay a solarization tarp over your beds 1 or 2 weeks before planting to warm the soil more quickly. The transparent plastic allows sunlight to pass through, warming the first inch or so (3 cm) of soil.

Maintenance

Weeding

If you are growing melons without a plastic mulch, we recommend hoeing the soil to loosen it, reduce weed pressure, and help irrigation water better percolate into the soil. During the summer, it's a good idea to spread a 3-inch (8 cm) layer of organic mulch (straw).

Pruning

1. Fruits form along the third generation of stems or vines. Regular pruning is needed to encourage plants to reach this stage. Once it has 4 healthy leaves, pinch off the end of the stem.

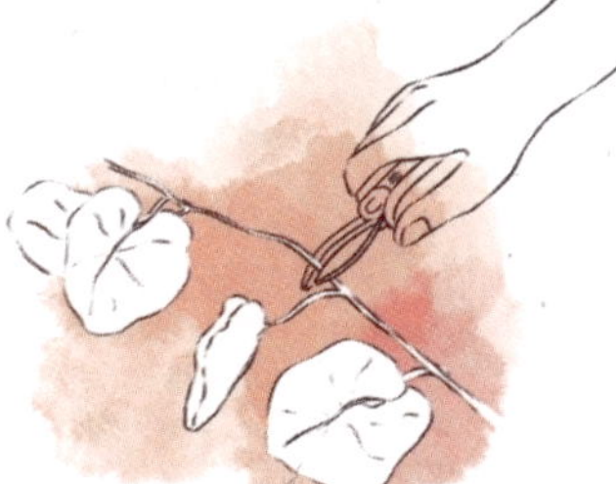

2. The plant will then grow two new stems. When these have 3 or 4 healthy leaves each, sever the stems just above a row of leaves, using sharp clean pruning shears.

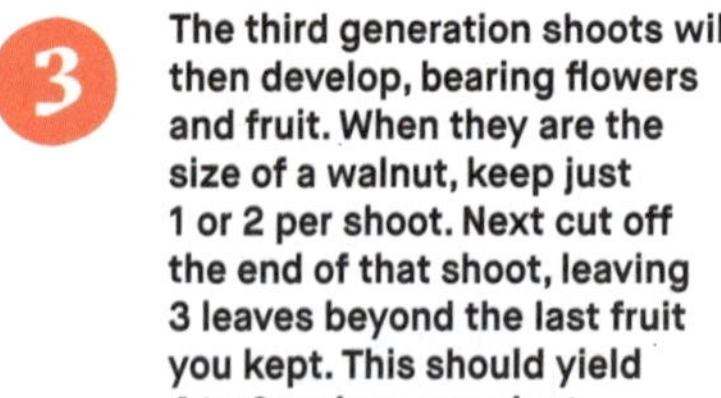

3. The third generation shoots will then develop, bearing flowers and fruit. When they are the size of a walnut, keep just 1 or 2 per shoot. Next cut off the end of that shoot, leaving 3 leaves beyond the last fruit you kept. This should yield 4 to 6 melons per plant.

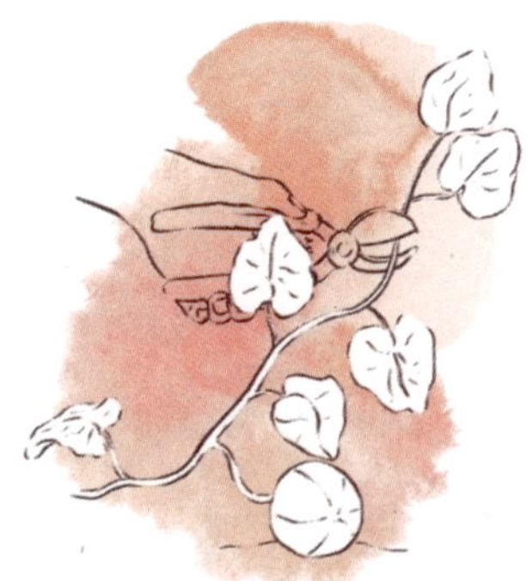

Fertilization

Melons are not very heavy feeders, so the manure you apply while preparing the bed is sufficient. However, once the plants bloom, about 6 weeks after planting, you can spray an algae-based foliar fertilizer containing trace elements and mineral salts. This will help plant development before fruiting begins.

Irrigation

Melon plants are known to be fairly drought resistant, and the foliage tends to wilt as plants dry out during the day. However, insufficient or irregular watering will slow fruit development, resulting in small melons. It's best to then use drip irrigation, reducing the water supply by a third, maybe even half, once the fruits are a bit bigger than a tennis ball to help boost sugar content and improve flavor.

Plant Protection

We highly recommend covering seedlings with insect netting to keep out aphids, thrips, and other pests. Watch out for slugs and snails that devour young leaves, and beware of rodents that feed on the fruit. When flowers form, remove the netting so that insect pollinators have access. Several diseases can emerge and cause damage, such as late blight, powdery mildew, anthracnose, fusarium wilt, verticillium wilt, leaf mold, and botrytis. Opt for varieties that are disease resistant and, above all, make sure your greenhouses are well ventilated to limit humidity levels that can foster these diseases.

Harvest

Picking

Melons must be harvested when they are perfectly ripe. If the fruit is too green, it won't be sweet enough; if it's too ripe, the flesh will be mushy and taste fermented. Several indicators can help you decide when it is ripe. With the Charentais melon, the skin changes from pale green to yellow. For the Cantaloupe melon, the stem separates from the fruit and the skin changes from grayish-green to yellow. When melons are ripe, they become highly fragrant. Use pruning shears to cut the stem, leaving about a 0.5 inch (1 cm) on the fruit, then carefully store the crop in harvest crates or bins. Melons should be eaten fairly soon after harvest.

Storage

Professional market gardeners can store melons in bins with lids until they are sold. Store them for no more than 4 days in a cold room kept at 36°F (2°C).

For home gardeners, we recommend eating the melons soon after harvesting to avoid refrigeration, which impacts aroma and flavor. Some varieties can be stored for a long time in a cool room and still maintain their flavor!

Focus on the Pepino Melon

The pepino melon, also called pepino, pepino dulce, melon pear, and melon shrub *(Solanum muricatum)*, produces white fruits with purple striping. Originally cultivated in South America, it was brought to Europe in the 19th century by British navigator and explorer Samuel Wallis.

A perennial plant cultivated as an annual, the pepino melon fruits have a texture similar to pears, with a melon-like scent. They are rich in vitamin C and trace elements (manganese and phosphorus) and can be consumed raw in juices and fruit salads or cooked into compotes, jams, and pies.

This plant does well in warm temperate climates and thrives when grown under cover. Like its cousin the eggplant, pepino melon needs rich loose soil and a sunny location.

Sow this crop in a greenhouse or bright heated room at the end of February. Pot up plants once they have 2 or 3 healthy leaves. Transplant seedlings into the ground after the last frosts, in one row spaced 24 inches (60 cm) apart in all directions.

In terms of fertilization and pruning, pepino melons are like tomatoes (we recommend removing suckers as they appear). Water sparingly and avoid overwatering.

Harvest about 6 months after seeding, 4 months after planting, once the fruits are ripe and feel slightly soft. Store them for several weeks in a cool room.

Watermelons

Native to Africa, watermelons are quite similar to other melons and cucumber, but they are much larger, and the flesh, which is less sweet, has a higher water content. These annual heat-loving tropical and subtropical plants have round to oval fruits with a tender, very juicy red or white flesh. The rambling vines, covered in white hairs, can reach 30 feet (10 m) long. Leaves are downy and palmate, with 3 to 5 lobes, and yellow flowers bear both male and female organs on each plant. Female flowers develop at the end of a thick stem that resembles a small fruit. Mostly made up of water, fruits have antioxidant properties and high levels of sugars and amino acids, promoting good cardiovascular health. Depending on the variety, they can be eaten raw, added to fruit salads, jams, and sorbets.

Watermelons

COMMON NAMES: Watermelon.
SCIENTIFIC NAME: *Citrullus vulgaris.*
FAMILY: Cucurbitaceae.
REQUIREMENTS: Full sun and warm weather with well-drained, rich, cool, deep sandy soil.
SPACING: 1 row down the middle of the bed with plants set 30 inches (75 cm) apart.
SEEDING: Professional growers can sow trays on heating mats in a greenhouse from mid-April; home gardeners can grow seedlings in a bright heated room, such as a sunroom.
TRANSPLANTING: Transplant seedlings 2 to 4 weeks after sowing (mid-May below the 45th parallel or end of May; in a greenhouse or low tunnel, north of the 45th parallel).
DAYS TO MATURITY: 120 to 150 days, depending on the climate.
ENEMIES: Slugs, aphids, powdery mildew.
VARIETIES: Sugar Baby (dark red flesh, very sweet, 4.5–6.5 lb., 2–3 kg), Sugar Belle (sweet red flesh, 5.5–9 lb., 2.5–4 kg), Pauline (red flesh, 17–22 lb., 8–10 kg), Mini-Love (sweet, few seeds, 2–7 lb., 1–3 kg), Crimson Sweet (red flesh, up to 30 lb., 14kg).
A NOTE FROM JEAN-MARTIN FORTIER: Watermelons are easy to grow and require little care. They thrive in the heat and require proper irrigation to ensure plenty of water. To get the best possible flavor, time your harvest carefully.

Growing Watermelons

> **"Although their bulk and weight make handling and transport difficult, watermelons are a hit in the summer because they are so refreshing."**
>
> Jean-Martin Fortier

Planting

Seeding

Sow 1 seed per pot filled with potting mix and keep them in a warm bright location. Water regularly and don't allow the space to get too hot as high temperatures will cause the main stem to become leggy.
From mid-May, you can also direct seed in loose soil. Make a hole and add 3 seeds, fill it in, tamp the soil, and water. After germination, once plants have 2 healthy leaves, keep only the most vigorous seedlings.

Transplanting

Place seedlings grown under cover outside in partial shade for a few days to harden off before transplanting them into the field. Loosen the soil and dig deep holes every 3 feet (90–100 cm) down the middle. Add compost to them then insert the seedlings without burying the stem. Tamp the soil and thoroughly water the base of each plant. Market gardeners then install a drip irrigation system.

Tip from Jean-Martin Fortier

Above the 45th parallel, we recommend installing a floating row cover or a caterpillar tunnel over this crop from the start of cultivation until June.

Maintenance

Weeding

Hoe the bed regularly until the end of June to loosen the soil then cover with straw in early July to keep the soil cool, reduce irrigation needs, and protect the fruit from splashing soil.

Pruning

Watermelon plants don't need any special pruning, but they do produce a lot of foliage. Simply cut back any invasive stems that are not bearing fruit.

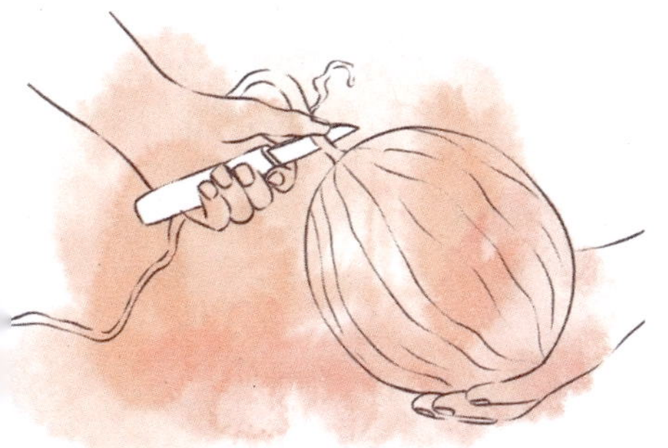

Plant Protection

When watermelon is direct seeded, the seedlings are vulnerable to slug damage. From late August to early September, watch for powdery mildew when the weather is wet or after irrigating with a sprinkler.

Harvest

Once a watermelon reaches its full size and maturity, the nearest tendril begins to wilt. The fruit should sound hollow when tapped. Another indicator is the yellow spot that sometimes appears where the fruit touches the ground.

Use a knife or pruning shears to cut the fruit from the stem and handle it with care.

Storage

Watermelons will keep for a few days when stored carefully in crates or on racks in a space maintained at 54°F (12°C).

Broad Beans (Fava)

Like peas and garden beans, broad beans belong to the Fabaceae family. They were domesticated around the Mediterranean as far back as 5000 BC and exported to China in the 1st century, then to Europe in the Middle Ages and the Americas in the 17th century. This annual species produces pods containing oval flattened seeds that have a creamy texture and a nutty taste. The pods of some varieties are also edible. The plants have thick, broad, oval, pointed leaves that range from grayish-green to dark green and produce large white flowers with black spots, which eventually turn into light green, pink, purple, or off-white seeds. Broad beans are rich in iron, calcium, phosphorus, magnesium, and potassium, and they are a significant source of protein. The seeds are eaten raw or cooked, mashed, in soups and stews, pan-fried, braised, or served as a side dish.

Broad Beans (Fava)

COMMON NAMES: Broad bean, fava bean, faba bean, bell bean, field bean, horsebean, Windsor bean, broad vetch.

SCIENTIFIC NAME: *Vicia faba.*

FAMILY: Fabaceae.

REQUIREMENTS: Sunny and warm aspect, deep but not heavy soil that warms up quickly in the spring.

SPACING: 1 row down the middle of the bed, or 2 rows set 14 inches (35 cm) apart.

SEED: Outdoors in October and November in regions with mild winters; elsewhere from February to mid-April.

DAYS TO MATURITY: 120 to 140 days, depending on the variety.

ENEMIES: Aphids, late blight, rust, broad bean weevils.

VARIETIES: Aguadulce (vigorous, productive, 8 or 9 seeds per pod), Witkiem (disease resistant for late blight, can withstand broad bean weevil damage), Seville (early and prolific), Karmazyn (dwarf, productive).

A NOTE FROM JEAN-MARTIN: This hardy legume heralds the arrival of spring vegetables and is easy to grow. However, as with all Fabaceae, harvesting it is a tedious task.

Growing Broad Beans

"In the early vegetable lineup, broad beans are a delight when harvested young and fresh—eaten raw, with a sprinkle of salt—while the seeds are still a little juicy and sweet."

Jean-Martin Fortier

Planting

Preparing the Soil

Loosen the soil with a broadfork and remove surface plant debris with a bed preparation rake. Then spread a few shovelfuls of sifted compost and use a tilther to incorporate the organic matter into the top 1.5 to 2 inches (4–5 cm) of soil. Level the bed with a rake.

Seeding

Use a rake fitted with line-marker tubes or use a rope pulled taut down the bed to mark out 1 or 2 rows, then dig furrows 2 to 3 inches (6–8 cm) deep with a hoe dag. If the soil is dry, lightly irrigate the bottom of the furrow with a trickle of water to moisten the seedbed.

Drop one seed roughly every 2 inches (5–6 cm) or sow 2 or 3 seeds per hole made every 5 inches (12–15 cm). Fill in the furrow and lightly tamp with the back of a rake. Water to keep the soil moist until germination, then irrigate very regularly until harvest as broad beans are sensitive to drought. Using a floating row cover to maintain temperatures of 46°F to 50°F (8–10°C) will accelerate germination and protect seedlings from early frosts (fall sowings) and late frosts (spring sowings).

Maintenance

Hilling

Once the plants are 12 inches (30 cm) tall, hill them with a rake or collinear hoe, covering a third of the stem length with soil. Use this opportunity to hoe and loosen the soil between the rows. Spread a straw mulch over the bed to reduce weed pressure.

Trellising

To keep the stems from bending under the weight of the pods, drive a line of 5-foot (1.5 m) stakes along the row, 12 inches (30 cm) into the ground and 10 feet (3 m) apart. Weave a double row of nylon lines, 6 to 8 inches (15–20 cm) above the ground, around the stakes and plants, connecting them in what is called the Florida weave. As the plants grow, place a second set of lines 8 to 12 inches (20–30 cm) higher if needed.

Harvest and Storage

Harvest fresh pods when they are still green and unopened, before seeds develop a mealy texture, roughly 100 days after sowing. Harvest dry beans when the pods are black. Fresh beans should be kept for no more than 2 to 3 days in a cold room, ideally 40°F (4°C).

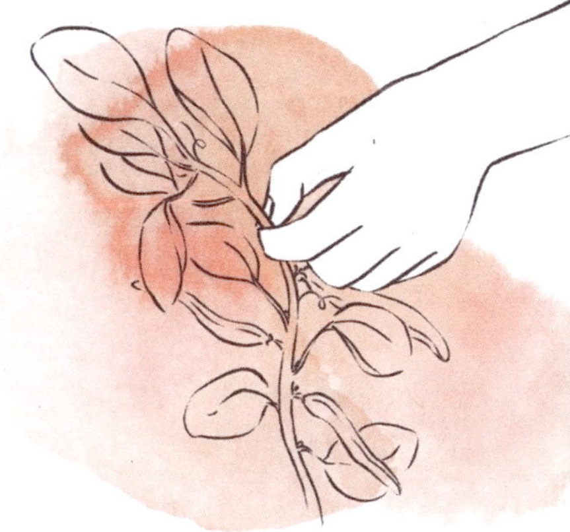

Tip from Jean-Martin Fortier

To prevent the stems from growing too high and to direct energy towards the pods rather than the foliage, pinch off the tips of the stems above the 5th or 6th row of well-developed beans.

Peas

Peas are annual plants, native to the Middle East, Egypt, and Ethiopia, where the seeds were eaten when ripe and dried. This is probably the world's oldest legume. It wasn't until the Middle Ages that people began to eat fresh peas, and widespread consumption only began in the 19th century, when certain varieties were selected for their delicate flavor. You can eat just the seeds or sometimes the entire pods, such as snow pea and snap pea varieties. The straight or slightly curved pods, green, purple, or yellow, contain smooth, wrinkled, or rounded seeds. Herbaceous, voluble, and branched stems grow to 20 to 24 inches (50–60 cm) tall for dwarf varieties and more than 6 feet (2 m) in climbing varieties. Leaves are made up of oval and opposite leaflets. Peas are a good source of protein and provide significant nutritional value due to their high starch, sugar, mineral, and vitamin content. Seeds can be eaten raw or half-cooked and crisp, or cooked with mixed vegetables, mashed, pan-fried, or in soups.

Peas

COMMON NAMES: Garden pea, field pea, dry pea, sweet pea, sugar pea, snap pea, snow pea, green pea.
SCIENTIFIC NAME: *Pisum sativum.*
FAMILY: Fabaceae.
REQUIREMENTS: Sunny and warm aspect, with light, loose soil that warms up quickly in the spring.
SPACING: 14 to 20 inches (35–50 cm), or 1 to 2 rows per 30-inch (75 cm) bed.
SEEDING: Direct seed in the fall in mild climates. Start under cover in February and seed until June, depending on the variety and the region.
DAYS TO MATURITY: 60 to 90 days, depending on the variety.
ENEMIES: Pea anthracnose, late blight, powdery mildew, pea rust, pea weevils, slugs and snails, pod borers, pea-leaf weevils, pea thrips.
VARIETIES: The many types of peas include those with round or wrinkled seeds enclosed in a cylindrical pod (Pois à Rames Téléphone, Serpette Verte de Malines) and snap and snow peas (Sugar Snap, Corne de Bélier Mangetout, Héraut Mangetout) with edible pods and seeds that are eaten fresh when young.
A NOTE FROM JEAN-MARTIN FORTIER: Peas are hardy and simple to grow. To make harvesting easier and to avoid having to bend down, opt for climbing varieties.

Growing Peas

"These vitamin-packed pearls are a big hit at the spring market, drawing in customers because there is nothing better than fresh peas!"

Jean-Martin Fortier

Planting

Seeding

In mild climates, sow peas in the fall under a high tunnel, caterpillar tunnel, or floating row cover. By the end of winter, once soil temperatures reach about 60°F (15°C), start spring seedlings with frost and cold protection.

1 **Loosen and level the soil using a bed preparation rake, spread 1 inch (2–3 cm) of compost over the bed, then incorporate it with a tilther.**

Tip from Jean-Martin Fortier

Seeds will germinate in 6 to 10 days if temperatures are 68°F to 72°F (20–22°C), but to help the germ to break through the seed coat, it's best to soften pea seeds by soaking them in a glass of water overnight before planting.

2

Mark rows using a bed preparation rake with plastic row markers slotted onto tines that are 14 inches (35 cm) apart. Next use the handle of a tool to dig a 1.5- to 2-inch (4–5 cm) furrow. Drop seeds every 1.5 inches (3–4 cm) or so and cover with a thin layer of soil.

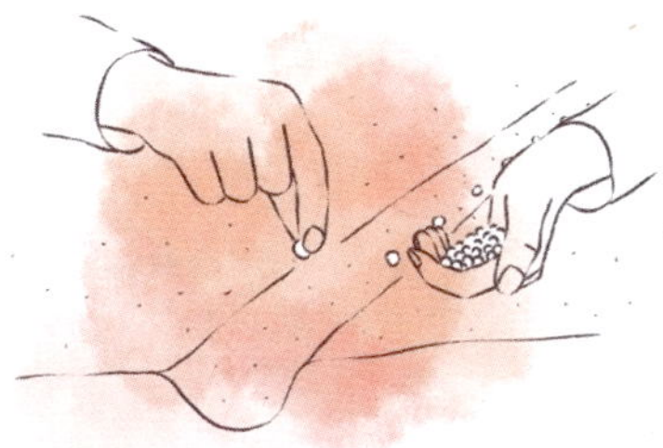

3

Water with a gentle spray until the crop germinates, then very frequently as peas are sensitive to drought. If you do not have a tunnel, protect the bed with a floating row cover, removing it after germination.

Maintenance

Hilling

When the pea plants are 6 to 8 inches (15–20 cm) tall, hill each row with a rake to promote rooting and support the stems, then spread pelleted chicken manure and alfalfa meal at the base of every plant.

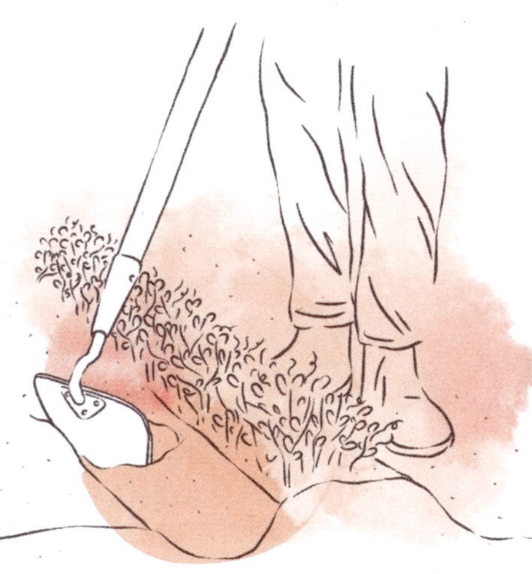

Mulching

After the plants have been hilled and are 12 inches (30 cm) tall, spread a layer of straw mulch to keep the soil cool and to reduce weed pressure. To facilitate watering, set up drip irrigation lines at the base of the plants before mulching.

Trellising

While climbing peas must be trellised, bush peas also benefit from support as the stems tend to fall over under the weight of the pods. This can be as simple as sticking branches into the ground for the plants to cling to or setting up two nylon lines tied to stakes that sandwich the plants between them.

Plant Protection

To prevent powdery mildew, which emerges in late summer, spray plants with a nettle tea. Apply a Bordeaux mixture fungicide to prevent late blight and anthracnose. Watch closely for both botrytis, which causes dark-brown leaf spots, and black aphids, which are very fond of tender Fabaceae stems.

Harvest

Picking

Harvest pea pods when they are still green, cylindrical, and tender. Using a harvest bag or a crate, start from the bottom of every plant, moving upwards as this crop ripens from the ground up. With one hand, grasp a pod and use your thumb and index finger to snap the stem just above the fruit. Transfer the pod to your other hand that is holding harvested peas, as you move along the row. Pick sugar and snap pea varieties when the fruits are still unripe and eat the entire pod raw and on the spot!

Storage

After picking peas, store the harvest in shallow crates so that air can circulate around the pods. If they are wet, store them in a cool, well-ventilated space.

Focus on the Asparagus Pea

The asparagus pea *(Tetragonolobus purpureus)* is another member of the Fabaceae family. In French, it is sometimes called *pois carré*, meaning "square pea" because the plant produces square-shaped pods, which taste like a cross between snap peas and asparagus. This vegetable, which has likely been cultivated since the 16th century, is extremely popular in India. The entire pod is edible and can be cooked like snow peas and string beans, best used soon after harvest. A high-calorie food, it is rich in calcium, magnesium, and vitamins A, B, and C and aids in digestion.

This bushy plant has no tendrils, grows to about 20 inches (50 cm) tall, and has a rather ornamental appearance. It thrives in full sun in deep, loose, cool, and humus-rich soil. With this plant, your biggest concerns should be late frosts and drought.

Sow seeds from mid-April in Mediterranean climates and from May to the end of June elsewhere. Watch out for slugs, which are particularly fond of this crop's tender stems and leaves! As with beans and peas, you must water regularly to keep the soil cool and moist. We recommend hilling to support the stems and applying compost or fertilizer (alfalfa meal or pelleted chicken manure) a few weeks before expected flowering.

After flowering, pods will grow quite fast, With harvesting from mid-June until the end of August. Pick pods when they are about 1 inch (2–3 cm) long.

Beans

The bean plant originated in the Andes, between Central and South America in the pre-Columbian era. Introduced to Europe in the 16th century, this annual is now one of the most widely cultivated vegetables in the world. There are several types, such as climbing pole beans, which are voluble plants, and bush beans. You can eat the pod and seeds or just the seeds, fresh or dried. Bush varieties grow no taller than 16 inches (40 cm) while pole beans can reach 10 feet (3 m). Large green leaves, made up of 3 leaflets, are pointed and oval; flowers are white, purplish, greenish, pink or lilac, depending on the variety. Green beans have diuretic properties and are rich in vitamins and mineral salts. Dried beans, a good source of energy, are rich in vitamin B, calcium, iron, phosphorus, and potassium. Cooked in butter, sauces and soups, sauteed...bean pods have many culinary uses. Dry beans can be cooked with parsley, baked, refried, braised, or added to salads, stews, soups, and cassoulet.

Beans

COMMON NAMES: Common bean, garden bean, runner bean, string bean.
SCIENTIFIC NAME: *Phaseolus vulgaris.*
FAMILY: Fabaceae.
REQUIREMENTS: Full sun. Healthy, loose, light, and humus-rich soil.
SPACING: For bush beans: 2 rows per bed, 20 inches (50 cm) apart, with 4-inch (10 cm) in-row spacing.
For pole beans: 1 row per bed, with 12-inch (30 cm) in-row spacing.
DIRECT SEEDING: In April below the 45th parallel and from May 15 to the end of July in more northern regions.
DAYS TO MATURITY: 60 to 70 days.
ENEMIES: Anthracnose, powdery mildew, halo blight, bean rust, bean weevils, spider mites, slugs and snails, bean seed flies, black bean aphids.
VARIETIES:
BUSH BEANS: Harvest every 2 days: Talisman, Fin de Bagnol, Triomphe de Farcy; every 4 days: Major Mangetout, La Victoire Mangetout, Mangetout Beurre Major (with mangetout varieties, the entire pod is edible).
POLE BEANS: Cobra, Emerite, Eva Mangetout, Helda Mangetout (with mangetout varieties, the entire pod is edible).
SHELLING BEANS: Haricot à Rames Coco Gros Sophie à Écosser, Haricot à Rames Langue de Feu à Écosser, Haricot à Rames Soissons Blanc à Écosser.
A NOTE FROM JEAN-MARTIN FORTIER: We only grow extra-thin beans, although harvesting is tedious and time-consuming. I feel it's important to grow this vegetable because it has unrivaled fresh flavor.

Growing Bush Beans

"Customers love beans, especially the thinner ones, which is why they should be grown on all micro-farms and in every veggie garden!"

Jean-Martin Fortier

Planting

Preparing the Soil

Beans grow well in carefully loosened soil that is level, well-drained, and warm. To ensure even germination, the soil temperature should ideally be at 68°F (20°C), definitely not below 60°F (15°C).

To warm up beds, you can cover them with a solarization tarp about 1 week before sowing. It may be wise to stagger your seeding dates to have longer and consistent harvests and avoid overproduction, especially in summer.

Seeding Rows

Before sowing, spread a 1-inch (2.5 cm) layer of compost and run a power harrow down the bed. Mark rows using a bed preparation rake fitted with plastic row markers slotted onto tines 14 inches (35 cm) apart.

2 Market gardeners can sow this crop using a Jang multi-row seeder. Home gardeners and professionals with a small-scale operation can use a single-row seeder or simply drop seeds by hand into furrows made with a garden tool handle.

3 After sowing, use a bed preparation rake to carefully cover up the rows and water with a gentle spray from a sprinkler or hose. Keep the soil moist for 7 to 10 days until the seeds germinate.

Cluster Sowing

1 On a bed that of loosened and leveled soil, mark 2 rows set 14 inches (35 cm) apart then dig holes every 8 inches (20 cm), dropping 5 or 6 seeds in each. Fill with soil, tamp, and water with a gentle spray.

2 When flowering begins, hill the soil up to the base of each plant with a hoe to keep the stems straight and to stop them from bending under the weight of the foliage.

Maintenance

Weeding

Young bean plants are vulnerable to weed pressure. The soil must be weeded and loosened with a collinear hoe every 10 days or so.

Irrigation

Home gardeners should water their crop with a watering can or hose, pointing the spray towards the base of the plant to avoid wetting the foliage. Professional market gardeners should set up drip tape to ensure regular irrigation at the base of each plant.

Fertilization

Two to 3 weeks after sowing, spread an organic fertilizer made from alfalfa meal and powdered poultry manure on both sides of the plants. Then use a garden claw to incorporate the organic matter into the soil and water thoroughly to dissolve and distribute it near the roots.

Plant Protection

Late blight thrives in cool, damp environments, but you can mitigate potential damage by choosing disease-resistant varieties.

In beans, anthracnose causes brown or even black spots on the leaves, as well as pink or gray spots on the pods. As a preventive measure, apply a horsetail tea solution (decoction). As soon as the disease appears, destroy affected plants and spray what remains of the crop with a Bordeaux mixture fungicide.

Tip From Jean-Martin Fortier

Once the plants are about 8 inches (20 cm) tall, we recommend applying a straw mulch between the rows to keep the soil moist and suppress weed growth. When it's time to hill the crop later, pull the straw away from the plants, then replace it.

Harvest

Picking

Every 2 days harvest beans that are thin, tender, without bulges. To pick faster, we recommend kneeling and collecting beans with both hands. Hold the pod with your index, middle, and ring fingers and use your thumb to snap the stem just above the fruit. Holding as many beans as you can before transferring them to a harvest bin saves considerable time.

While harvesting, be careful not to tug on the plant as you could easily pull up its shallow root system.

Storage

Fresh beans don't keep for long. Home gardeners can freeze them raw or cooked. After trimming the pods, dunk them in boiling water for 4 minutes, followed by an ice bath. Drain and spread them out on a dish towel to thoroughly dry before putting in airtight bags and placing in the freezer. Professional market gardeners should store unwashed beans in crates in a cold room set to 39°F (4°C). However, avoid storing for more than 4 days.

Focus on the Pole Bean

Pole beans, which deliver yields earlier and longer than bush beans, can climb up to 13 feet (4 m) and therefore must be trellised.

In greenhouses, support plants with nylon strings tied to the overhead structure. Some varieties have been bred specifically for greenhouse cultivation. Wherever you grow beans, expect to harvest them after 8 to 10 weeks.

Professional market gardeners should sow pole beans into plug flats on heating mats set to 86°F (30°C) in a greenhouse until germination. When seedlings are 4 to 6 inches (10–15 cm) tall, plant them in groups of 3 every 12 inches (30 cm) along the row.

Home gardeners can prepare the soil and install trellising before direct sowing pole beans as described for bush beans.

If growing outside, drive chestnut stakes into the ground and tie them together like a tripod or set in parallel rows and joined at the tops.

In a greenhouse, tie nylon lines to its overhead structure every 12 inches (30 cm) and use tomato clips to connect each plant to one line.

When trellising beans, be very careful because the stems are fragile.

Once pole beans reach nearly 10 feet (3 m) high, cut the stems (top the plants). Repeat this operation 2 or 3 times during the growing season to direct the plants' energy towards pod production.

Growing Shelling Beans

"Shelling beans are always a hit because they are just as delicious fresh as they are dry."

Jean-Martin Fortier

Planting

Preparing the Soil

Both types of dry beans, those that are eaten fresh and those that are eaten dry, thrive in light well-drained soil that is rich but not humic. These plants grow poorly in soils recently amended with compost, manure, and lime. Use a broadfork to loosen the soil down to 8 inches (20 cm) then run a tilther down the bed to level the surface and create a finer soil texture.

Seeding

Transplant seedlings from mid-April below the 45th parallel, or between mid-May and mid-July in more northern locations, much like sowing bush and pole beans. To direct sow in the field, make a furrow 1.5 inches (3–4 cm) deep and drop 4 or 5 seeds together every 16 inches (40 cm). Fill in the furrow and tamp with the back of a rake.

Caring for Seedlings

Water seeds with a gentle spray until germination then water regularly and more thoroughly during flowering. In cold climates, protect seedlings with a floating row cover and hill them when about 6 inches (15 cm) tall so they don't fall over. Regularly hoe the crop to keep weeds in check and break the soil crust caused by rainfall.

Trellising

Bush varieties bred for shelling do not require support. However, all types of dry beans (pole and bush) must be supported by stakes installed in either in a tripod or a row connected by a horizontal slat. This trellising structure should be about 8 feet (2.5 m) tall.

Tip from Jean-Martin Fortier

Instead of making a tripod, you can create a trellis by driving stakes into the ground every 6 feet (2 m), connecting them along their tops, and attaching plastic mesh or sheep fencing for the voluble bean stems to climb.

Maintenance

Supports

Guide young shoots towards the stakes or netting so they have something to cling to and grow taller. Once they have been trellised and the stems are 14 to 16 inches (35–40 cm) tall, hill the plants a second time, in dry weather, using a hoe, a rake, or a hoe dag, to anchor and support the root system.

Harvest

Eat fresh beans shortly after harvesting the well-formed pods. You can also harvest when the plants are dry and yellowing at the end of the summer, into September, always before the first frost. Pull all the plants and store in a dry and well-ventilated space to dry out. Store the whole plants or shell the beans and discard the shells, stems, and leaves.

Tip from Jean-Martin Fortier

Because pods are vulnerable to bean weevil damage, you must store them in airtight containers. Put dry beans in the refrigerator or a cold room for 24 hours before storing them to kill any remaining larvae.

Tomatoes

Native to South America, tomatoes were cultivated by the Aztecs as early as 500 BC. Spanish conquistadors brought them to Europe where they were initially considered to be ornamental. In the 18th century, European market gardeners contributed to the growing popularity of this member of the Solanaceae family, now grown in 170 countries. While the plants are perennial in tropical climates, they are grown as annuals in more temperate regions. Fruits can be red, orange, black, and yellow, with striping in various hues and come in all shapes and sizes. They are classified based on weight: currant tomatoes (less than 0.4 oz., 10 g), cherry tomatoes (0.4–0.7 oz., 10–20 g), cocktail tomatoes (0.7–1.4 oz., 20–40 g), medium-sized tomatoes (1.4–3.5 oz., 40–100 g), large tomatoes (3.5–10.5 oz., 100–300 g), and beefsteak tomatoes (over 10.5 oz., 300 g). High in fiber and low in calories, all contain carotenoids (lycopene) that are known for their antioxidant and anti-inflammatory properties. They are also rich in minerals and vitamins. Raw or cooked, tomatoes are an essential ingredient in cuisines around the world.

Tomatoes

COMMON NAME: Tomato.
SCIENTIFIC NAME: *Solanum lycopersicum.*
FAMILY: Solanaceae.
REQUIREMENTS: Tomatoes thrive in warm temperate climates, consistent temperatures between 64°F and 81°F (18–27°C), and sunny locations sheltered from the wind. In northern climates, they are grown under cover.
SPACING: Under cover, plant every 2 feet (60 cm) in 2 staggered rows 30 to 40 inches (80–100 cm) apart. Plant field tomatoes every 2 feet (60 cm) in a single row.
SEEDING: In March in a bright, heated room, such as a sunroom, for home gardeners. Market gardeners can stagger seedings (sow successions) from late January to April in a greenhouse.
TRANSPLANTING: From mid-April to mid-May under cover and in May for field tomatoes.
DAYS TO MATURITY: 120 to 150 days.
ENEMIES: Late blight, early blight, gray mold or botrytis, powdery mildew, and verticillium wilt. Cutworms (moth caterpillars), whiteflies, slugs, nematodes, and tomato leaf miners *(Tuta absoluta)*. Physiological issues (blossom-end rot, puffiness, discoloration, or cracking).
VARIETIES: There are more than 2,700 kinds of tomatoes around the world, including determinate and indeterminate varieties.
Determinate tomatoes (bush): These crops are suited to temperate climates, produce small fruits, require no pruning, and are ideal for small spaces.
Indeterminate tomatoes (indefinite growth): These varieties grow continuously until they die due to frost or disease.
A NOTE FROM JEAN-MARTIN FORTIER: Tomatoes are the quintessential summer crop that every professional and home gardener should grow, and it's virtually foolproof. For more information, read *Tomatoes: A Grower's Guide* in the Grower's Guides from the Market Gardener series.

Cherry and Cocktail Tomatoes

Apero F1

Apero F1 is a hybrid cherry tomato that is ideal for appetizers, aperitifs, or cocktails, as its French name suggests.

DESCRIPTION: Highly productive, with uniform fruits. Consistent, long fishbone trusses.

A NOTE FROM JEAN-MARTIN FORTIER: Highly disease resistant. Rich flavor, very thin skin. Sweet and delicate taste.

Black Cherry

One of the best cherry tomatoes, this variety was bred by Vince Sapp for the Tomato Growers' Supply.

DESCRIPTION: Productive until late summer or early fall, 10 to 12 fruits per cluster, brown flesh. Ripens well after being harvested while green.

A NOTE FROM JEAN-MARTIN FORTIER: Well adapted to warm regions, cool climates, and short seasons. No pruning required. Fairly disease resistant. Not prone to cracking. Juicy, sweet, and tangy.

Clementine

Clementine is another appetizer classic that brings together both flavor and color!

DESCRIPTION: Fruit the size of a gooseberry or small plum. Highly productive, about 100 tomatoes per plant, uniform in size.

A NOTE FROM JEAN-MARTIN FORTIER: Adapts well to cooler climates. No pruning required. Juicy, sweet, and very flavorful.

Auriga

This variety was first distributed by Dr. Martin Stein in the former East Germany and later rediscovered by the Swiss after the fall of the Berlin Wall.

DESCRIPTION: Very hardy. Consistent, high-quality yields starting mid-season. Large quantity of trusses of 4 to 7 tomatoes. Smooth skin, slightly elongated shape. High carotene content.

A NOTE FROM JEAN- MARTIN FORTIER: Vigorous and hardy. Suitable for cold regions. Not prone to cracking. Firm and juicy, with a thick skin. Rich and sweet flavor. Use in salads and sauces.

Medium to Large Fruits

Marnero F1

Marnero F1 is a hybrid tomato bred from the Marmande variety for improved disease resistance.

DESCRIPTION: Highly productive. Uniform fruits. Dense flesh. Tops of fruit (shoulders) are darker and often green. Contains very few seeds.

A NOTE FROM JEAN-MARTIN FORTIER: Good pest and disease resistance. Sweet, with complex flavor. Use in salads, cooked, and stuffed.

Rose de Berne

This Swiss heirloom variety is quite famous in France and is considered one of the best tomatoes because of its flavor.

DESCRIPTION: Rose de Berne produces clusters of 4 to 6 tomatoes. Small yields, but flavorful, dense, juicy, and sweet.

A NOTE FROM JEAN-MARTIN FORTIER: This is one of the best tomatoes for salads. The skin is very thin, while the flesh is dense, juicy, sweet, and fragrant.

Marmande

Developed by Vilmorin & Cie in the 1930s, this variety is a showstopper in vegetable gardens.

DESCRIPTION: Vigorous, irregular-shaped fruits, highly productive. Adapts well to cold climates and short seasons, which explains its success in gardens.

A NOTE FROM JEAN-MARTIN FORTIER: Marmande tomatoes have a typical heirloom flavor, complex, with high sugar content. They are versatile, and can be either raw or cooked.

Brandywine

With irregularly shaped fruits, Brandywine has frequently been used in crosses—intentional (Purple Brandy) and accidental (Lucky Cross).

DESCRIPTION: Round, slightly flattened, and ribbed. Particularly abundant harvests in the second half of the season.

A NOTE FROM JEAN-MARTIN FORTIER: For healthy growth and to avoid plants getting long, thin, and pale, provide good exposure to light and sun. Juicy and dense flesh, remarkable vine-ripened flavor and aroma. Versatile uses: raw in salads; cooked in sauces and coulis.

Heirloom and Beefsteak Tomatoes

Pineapple

This heirloom variety was revived in the 1950s and remains one of the best from a flavor standpoint.

DESCRIPTION: Good yields. Skin and flesh are the same color. Shoulders are quite wide and sometimes not mature.

A NOTE FROM JEAN-MARTIN FORTIER: Better suited to warm sunny regions. Sweet complex flavor, not very juicy. Use in salads and carpaccio. Loses some flavor when cooked.

Hawaiian Pineapple

Of unknown origin, this tomato was first marketed in the 1970s in southern Indiana.

DESCRIPTION: Highly productive, large beefsteak variety. Irregular ribbed shape. Orange flesh with red streaks.

A NOTE FROM JEAN-MARTIN FORTIER: Very fruity and sweet, reminiscent of pineapple when ripe.

Beefsteak

This American tomato is similar to Coeur de Boeuf (Cuor di Bue, Oxheart), which has been bred into many other varieties.

DESCRIPTION: Produces very large, round flattened tomatoes. Slices look meaty, hence the name. Contains few seeds.

A NOTE FROM JEAN-MARTIN FORTIER: With a meaty melt-in-your-mouth texture, this variety can be eaten in salads or stuffed. It has a sweet complex flavor.

Gregori's Altai

This variety comes from the Altai Mountains on the border between Russia and China.

DESCRIPTION: Produces deep pink tomatoes weighing 7 to 17.5 ounces (200–500 g). Highly productive and well adapted to short seasons and warm climates.

A NOTE FROM JEAN-MARTIN FORTIER: This variety is quite popular for its sweet flavor and dense flesh that can be thinly sliced, ideal for carpaccio.

Original and Colorful Tomatoes

Black from Tula

This variety, originally from Tula, Mexico, was mentioned by Carolyn J. Male in her book, *100 Heirloom Tomatoes for the American Garden.*

DESCRIPTION: Productive. Green shoulders, dark red flesh. Slight ribbing at the shoulders.

A NOTE FROM JEAN-MARTIN FORTIER: Good mid-season yields. Sweet and highly flavorful, with meaty flesh. Use in salads.

Jaune Saint Vincent

This heirloom variety from France is one of the best yellow tomatoes.

DESCRIPTION: Highly productive. Smooth skin, distinct ribbing at the shoulders.

A NOTE FROM JEAN-MARTIN FORTIER: Well adapted to short seasons and cooler climates. Juicy, sweet, mild, and tangy. Quickly becomes mealy once ripe. Use in salads or carpaccio.

Black Krim

This heirloom variety from the Black Sea region of Crimea is extremely popular with tomato lovers.

DESCRIPTION: Prolific producer. Dense flesh. Shoulders are darker and often green. Contains few seeds.

A NOTE FROM JEAN-MARTIN FORTIER: Good drought resistance, not prone to disease. Does not tolerate excessive watering (splitting). Sweet, delicate flavor, nonacidic. Use in salads.

Coeur de Boeuf

The real Coeur de Boeuf (*Cuor di bue* in Italian, meaning "ox heart") is heart-shaped as its name suggests. It has often been imitated but never equaled! Distinctive tomato flavor.

DESCRIPTION: Productive, contains few seeds, and has been bred into many different varieties, resulting in several colors.

A NOTE FROM JEAN-MARTIN FORTIER: Prune misshapen and excess fruits to promote the development of a few exceptional specimens. Very sweet, with dense flesh and rich flavor. Use in carpaccio and salads.

Growing Tomatoes

"The tomato is without a doubt a signature crop for any good market gardener."

Jean-Martin Fortier

Planting

Seeding into Plug Flats

1. Fill plug flats with moistened potting mix, level by hand, and drop one seed into each cell. Cover with a thin layer of soil and tamp lightly.

2. Water with a gentle spray and place flats in a greenhouse on heating mats set to 75°F (24°C) until germination. Once seeds sprout, remove the heating mats and maintain ambient temperature at 75°F (24°C) during the day and at least 64°F (18°C) overnight.

Seeding into Open Flats

In a flat with a perforated bottom, spread a layer of potting mix and sow with a small hand seeder or use the seed packet edge as a spout.

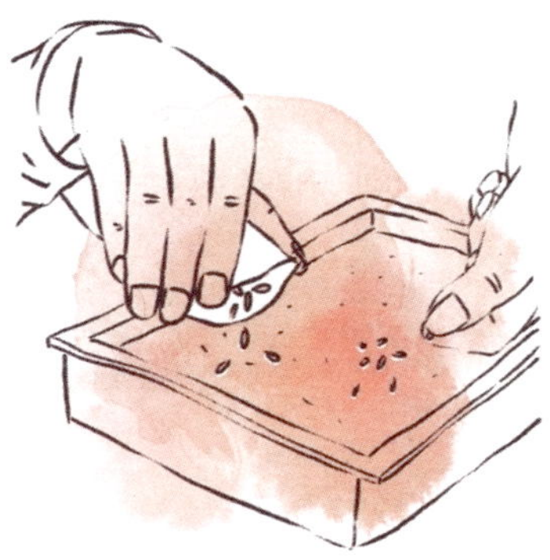

2

Cover seeds with a thin layer of sifted potting mix, tamp, and water with a light spray. Leave the tray in a bright warm space. Water sparingly until germination then more often afterwards.

Potting Up

After about 3 to 4 weeks, seedlings form 2 leaves that are 4 to 6 inches (10–15 cm) long. At this point, tomatoes sown into open flats or plug flats start running out of space, but they are still too young and fragile to transplant. In this intermediate stage, transfer each seedling to a pot roughly 4 inches (10cm) wide.

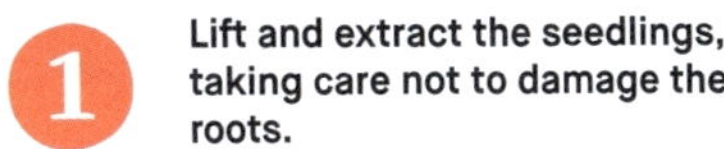

Lift and extract the seedlings, taking care not to damage the roots.

2 Fill 3- to 3½-inch (8–9 cm) pots with potting mix, then dig a hole that is wider than the root ball, leaving about an inch (a few centimeters) of soil depth.

Bury seedlings up to the cotyledons (first row of leaves). Gently tamp the soil without compacting it.

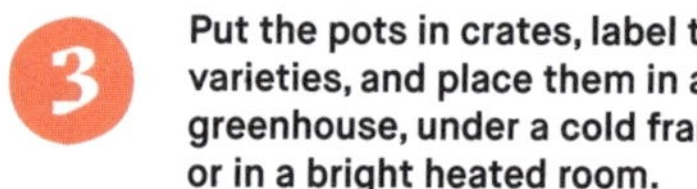

3 Put the pots in crates, label the varieties, and place them in a greenhouse, under a cold frame, or in a bright heated room.

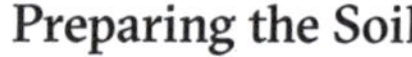

Preparing the Soil

Tomatoes need rich, loose soil grown under cover, in a greenhouse or tunnel, or in the field. Lay a silage tarp (occultation tarp) over the beds a few weeks before planting to kill weeds that grew over the winter and to benefit from earthworms loosening the soil.

1 Loosen the soil with a broadfork and spread a 2-inch (5 cm) layer of compost.

2 Using a bed preparation rake, level the soil surface while incorporating the compost into the soil.

Tip from Jean-Martin Fortier

To prepare tomato seedlings for growing conditions in beds, whether under cover or in the field, we recommend hardening them off by putting them outside under an awning, exposing them to cooler temperatures while still sheltered from wind and sunlight. In this stage, you can add a pinch of nitrogen-rich chicken manure to each cell.

Use a tilther set to a depth of 2 inches (5 cm) to loosen the surface and create a finer soil texture that will promote better root development.

Tip from Jean-Martin Fortier

In tunnels and greenhouses, you can apply a woven ground cover as mulch. Just before transplanting, use a blow torch or hole burning tool to make 2 staggered rows of 5-inch (12 cm) holes every 24 inches (60 cm) in the fabric. Put one tomato seedling in each hole.

Transplanting

Plant tomatoes between mid-April and mid-May, depending on your region's climate. Crops grown under cover (low tunnel, greenhouse) tend to be planted more densely than in the field so use double-row cultivation methods. Growing tomatoes in the field is more common for home gardeners who cannot invest in a greenhouse or tunnel.

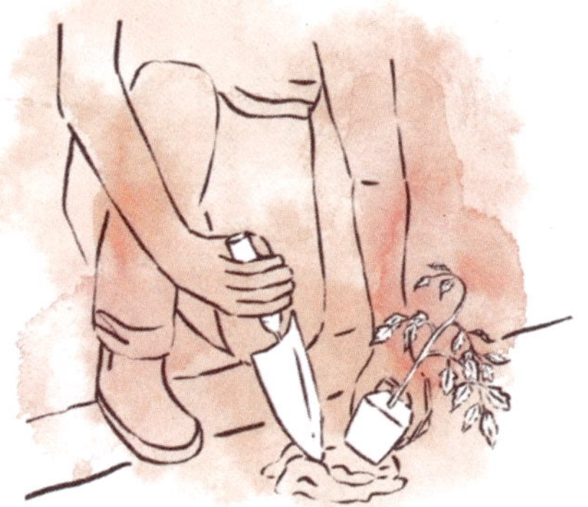

1. **Dig a hole with a garden trowel, remove each seedling from its pot, and put the root ball and stem in the hole. Fill in the space around the root ball and gently tamp the soil.**

2. **Water thoroughly at the base of each plant to settle the soil around the roots, topping with more soil if necessary. Cover the ground with a 3- to 4-inch (8–10 cm) layer of organic mulch, such as straw or coarse compost.**

Tip from Jean-Martin Fortier

Burying the first 2 to 4 inches (5–10 cm) of the stem promotes additional root development, which improves plant growth and vigor. If nights are still cool in the field, wrap a protective cover around young plants to create a warm microclimate around foliage and flower buds.

Maintenance

Trellising Under Cover

Without support, tomato plants tend to collapse under the weight of their fruit, eventually growing along the ground, exposing them to diseases such as late blight. It is therefore essential to tie the stems to sturdy supports immediately after transplanting. There are many established ways to stake tomatoes, depending on how they are grown. In tunnels and greenhouses, growers usually trellis tomato plants with nylon lines directly connected to the structures.

1. **Above each tomato plant, tie one nylon line to the overhead structure of the shelter, allowing it to reach the ground. Then loosely tie it around the base of the main stem.**

2. **As the plant grows, use plastic tomato clips to connect the stems, leaving room for them to grow.**

Training Field and Garden Tomatoes

Stakes can be made from various materials (wood, plastic, or metal), but they must be sturdy and weather-resistant.

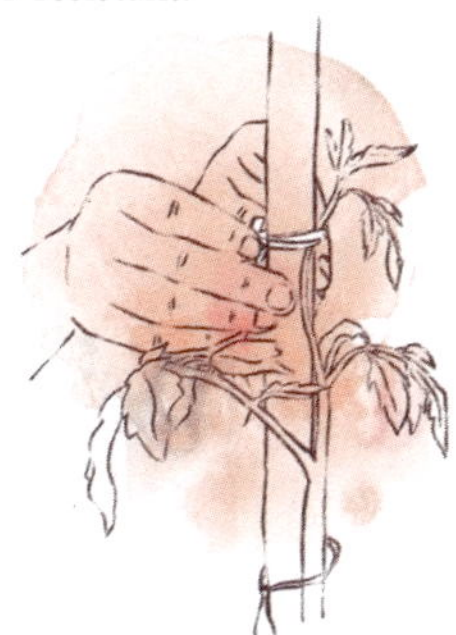

1. **Drive each stake 1 foot (30 cm) deep beside the seedling right after transplanting or a few days later, making sure not to damage the roots.**

2. **Using cotton string or twine, loosely tie the base of the main stem to the stake, leaving room for the stem to thicken and move along the stake as the soil settles. As the tomato plant grows, fasten the stem to the stake every 10 to 12 inches (25–30 cm).**

Tip from Jean-Martin Fortier

You can also use horizontal trellising to strengthen staking. Drive 2 chestnut stakes into the ground, with 4 tomato plants between them, then connect the tops with wire or wooden slat. Tie a line above each plant and attach it to the stem at ground level. As the seedlings grow, wrap the strings around the stems.

Irrigation

After planting the seedlings or just before the first watering, build a watering basin, a berm 4 to 8 inches (10–20 cm) high, around each plant. This keeps the water near the plant, prevents run off, and allows the water to percolate, making it directly accessible to the roots.

Drip irrigation delivers water right at the base of each plant. Adding a timer to the system ensures accurate and efficient water.

Pruning

Pruning is a time-consuming but necessary part of tomato care if you want to maximize yields, harvest good-quality fruits, and prevent the spread of disease. Only prune indeterminate varieties that produce large fruits. Determinate varieties, those with smaller fruits, and micro dwarf plants require less pruning, if any at all. The two essential pruning steps are removing suckers and removing lower leaves.

Pruning Suckers

Once a week, prune side shoots (suckers) growing from leaf axils by snapping them off with your thumb and index finger. Remove bigger suckers using a knife or shears with disinfected blades.

Pruning Lower Leaves

When tomatoes begin to ripen, remove lower leaves to limit the spread of soil-borne cryptogamic diseases (fungal, algal) and allow sunlight to improve the flavor and coloring. First remove the bottom leaves then work your way up to the first fruit cluster. Repeat this operation every 2 weeks as soon as summer begins.

Fertilization

During the foliar growth stage, apply an organic fertilizer (chicken manure, guano, horn meal). Later use a fertilizer with a high potassium content to promote fruiting. To boost the plants' natural defenses, apply nettle or comfrey tea solutions every 2 weeks while watering.

Plant Protection

Nutrient deficiencies and diseases can show up in many different ways. The various solutions are described in *Tomatoes: A Grower's Guide* in this book series, which is dedicated entirely to tomatoes.

Tip from Jean-Martin Fortier

Tomatoes need about 1 quart (1 liter) of water per day, an amount that depends on their stage of development, the growing method (under cover or not), and the soil quality. Ideally, your first watering should happen 3 hours after sunrise, and a second 4 hours before sunset. Avoid irrigating in full sun and especially do not overwater or underwater as tomato plants are highly sensitive to inconsistent irrigation, which can cause blossom-end rot and cracking.

Harvest

Picking

Tomato maturity can be divided into two stages: ready to pick and ready to eat. If you want to store tomatoes for a few days before eating them, harvest them at the ready-to-pick stage, still a little firm and having some color. However, if you plan to eat them immediately, pick when the flesh is soft and skins have deeper coloring. With cherry tomatoes, simply pull the fruit to separate it from the truss. With other varieties, gently grasp the fruit with one hand and use your thumb and forefinger to snap the stem right above it. Pruning shears make it easier and quicker to cut without pulling on the fruit and potentially damaging the truss. Place tomatoes side by side in a basket or crate in a single layer.

Storage

If you cannot eat tomatoes right away, make them into a sauce, dehydrate them, or simply cook and store them in containers. At the end of the season, harvest the last green tomatoes and let them ripen gradually. However, they will have slightly less flavor.

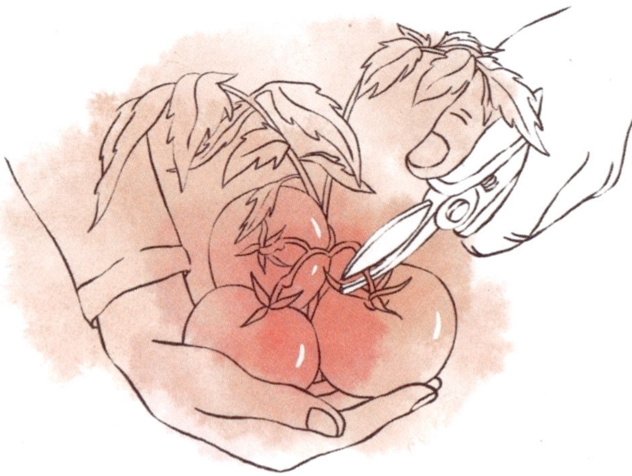

Tip From Jean-Martin Fortier

At the end of the season, if any plants still have underdeveloped fruits, pull them out and hang in a warm room (sunroom, tunnel) to be eaten gradually after they ripen.

Focus on the Sticky Nightshade

Sticky nightshade *(Solanum sisymbriifolium)* belongs to the same family as tomatoes and bears fruits that resemble cherry tomatoes. Sometimes called a litchi tomato, the plant is native to Central America and forms a bush with prickly stems.

It has fairly large green leaves that give it an ornamental appearance. Flowers, white or bluish and star-shaped, resemble eggplant and potato flowers and bloom in late summer. The vermilion fruits grow inside a prickly calyx, or husk.

Sticky nightshade thrives in warm and sunny locations and adapts to all soil types. In April, start seedlings in a greenhouse or warm room, to be transplanted after the first frosts.

Harvest 5 to 6 months later, from September until the first frosts. We recommend wearing gloves to protect against thorns.

Fruits are sweet and tangy with aromas of sour cherry, tomato, and lychee. This small uncommon fruit has many uses—in juices, fruit salads, appetizers, and jams—and is sure to delight food lovers seeking novelty.

Physalis
(Cape Gooseberry, Peruvian Ground Cherry, and Others)

A member of the Solanaceae family (like potatoes, tomatoes, eggplants, and peppers), the *Physalis* genus includes the Cape gooseberry, Peruvian ground cherry, tomatillo, and Chinese lantern. Cape gooseberry and ground cherry are grown for their fruit, a yellow-orange edible berry that is the size of a cherry tomato. It grows inside a lantern-shaped papery calyx that dries out and opens up when mature. The Greek name *phusallís*, meaning "bladder," refers to this shape. Native to South America, likely Peru, this voluble annual plant can reach over 6 feet (2 m) tall. Branches bear heart-shaped green leaves, and creamy-white flowers grow from the leaf axils in July. Their berries contain many vitamins (A, B, C, D, K) and properties that reduce joint pain, increase urination, promote relaxation, and relieve constipation. The tender fruits evoke flavors of gooseberry, tomato, and sometimes mango. They can be eaten raw and fresh in salads and on pastries or cooked into jams and pies.

Cape Gooseberries, Peruvian Ground Cherries, and other Physalis varieties

COMMON NAMES: Cape gooseberry, Peruvian ground cherry, golden berry, Inca berry, Aztec berry, giant ground cherry.

SCIENTIFIC NAME: *Physalis edulis, Physalis peruviana,* and other species in the *Physalis* genus.

FAMILY: Solanaceae.

REQUIREMENTS: These are heat-loving plants! They can be grown under cover (caterpillar tunnel) or in the field.

SPACING: 1 row per bed (30-inch or 75 cm bed width) with an in-row spacing of about 24 inches (60 cm).

SEEDING: From March to the end of May, either in a mini greenhouse, a sunroom, or a bright and warm room, or in a greenhouse on a heating mat.

TRANSPLANTING: In May as soon as possible after the last frost.

DAYS TO MATURITY: 150 days.

ENEMIES: Caterpillars, whiteflies, rust, and leaf spot.

VARIETIES: This genus includes several small fruits with husks, such as Cape gooseberry, ground cherry, tomatillo, and Chinese lantern. We recommend Peruvian Ground Cherry *(Physalis edulis)*, Dwarf Ground Cherry *(Physalis pruinosa)*, Caged Love Berry *(Physalis alkekengi*, grown as an ornamental), Mexican Tomatillo *(Physalis ixocarpa)* (see p. 99), and Ground Cherry *(Physalis pubescens*, also forms a ground cover, produces tart fruit).

A NOTE FROM JEAN-MARTIN FORTIER: Physalis are grown in the summer and don't require much attention. They are hardy and can become invasive. Trellising helps control them and makes harvesting easier.

Growing Physalis Species

"At the market, Cape gooseberries and Peruvian ground cherries attract attention with their lantern-shaped husks, while on plates, the colorful berries deliver surprising and subtle flavors!"

Jean-Martin Fortier

Planting

Seeding

Professional market gardeners and home gardeners can sow this crop in pots, open flats, or plug flats at a depth of ⅛ inch (3 mm). Cover the seeds with a thin layer of sifted potting mix then water with a gentle spray before placing in a greenhouse or warm and bright room. Keep the soil moist until germination, which happens quickly if ambient temperatures are around 72°F to 75°F (22–24°C).

Potting Up

When seedlings have 3 or 4 well-developed leaves, transplant into 4-inch (10 cm) pots. Gently pull the plants out with your fingers and a stick or chopstick, taking care not to damage the roots. Make a hole in the soil big enough to insert the root ball, gently tamp, and water. They can be transplanted into the field 4 to 5 weeks later.

Transplanting

Loosen the bed and amend the soil with a layer of compost and pelleted chicken manure. Run a tilther down the bed at a depth of 2 inches (5 cm) to level the surface and create a finer soil texture.

2 Pull a string taut down the bed then use a dibber to dig holes every 24 inches (60 cm). Before planting seedlings, make sure to water them so the potting mix is moistened. Bury the seedlings until just below the first leaves to promote better root development. Fill in the hole, firm down and water the soil.

3 Cover the bed surface with 2 to 3 inches (5–8 cm) of straw mulch and install one stake per plant. Market gardeners can drive stakes into the ground every 4 plants, with a slat connecting them along the tops. Tie a nylon string to the slat to support each main stem.

Tip from Jean-Martin Fortier

Direct sow Cape gooseberries and ground cherries from May to June when the temperatures are around 64°F to 68°F (18–20°C). Sow multiple seeds every 35 inches (90 cm) in a single row. This crop germinates slowly, about 3 weeks after sowing. When plants have 3 or 4 leaves, thin and keep only the strongest seedlings. After the crop flowers, irrigate regularly, directing water at the base of each plant to keep the foliage dry and prevent disease. Harvest in late August through October.

Maintenance

Weeding

If the soil is not mulched, weed and aerate the surface with a collinear hoe between the rows. Pull out any remaining weeds by hand.

Watering and Pruning

After the crop has flowered, water regularly in dry and warm weather and prune the tips of overgrown vines to help berries grow larger.

Plant Protection

While quite hardy, plants can be damaged by whiteflies, which happens with most Solanaceae. Apply a black soap solution or spray with 1 quart (1 liter) of water containing 20 drops of rose geranium essential oil. Remove any potato beetles by hand. As a last resort, spray a biopesticide.

Harvest

Picking

Ripe fruits turn golden-yellow in August while the calyx becomes pale and papery. On a sunny day, harvest by gently grasping the stem just above the fruit, using your thumb and index finger. Be careful not to damage the husk.

Storage

Store harvests in crates, layered but not too high to avoid crushing the fruits. They will keep for several weeks in a well-ventilated room at 55°F (13°C).

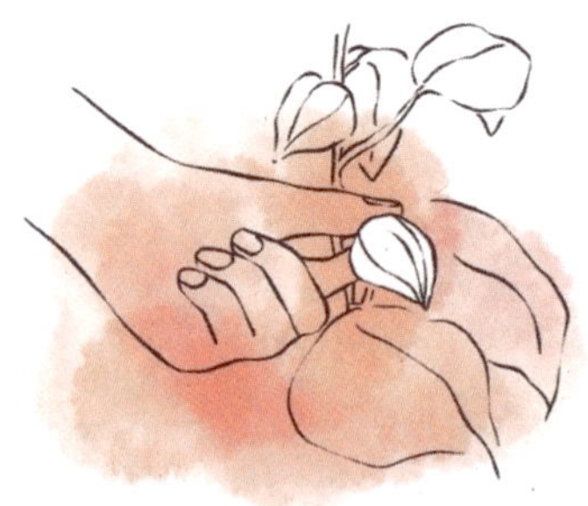

Focus on the Tomatillo

Tomatillo *(Physalis ixocarpa)*, with varieties such as Toma Verde, Amarylla, Purple Xtrem, and Violet, is a cousin of both Cape gooseberries *(Physalis peruviana)* and tomatoes. Native to Central America, this perennial plant is grown as an annual bush roughly 3 feet (1 m) tall with fruits that resemble cherry tomatoes.

With a tangy flavor, the fruits are mainly used in Mexican dishes, such as the legendary salsa verde that is eaten with tacos and enchiladas. They can also be served raw, in salads, or cooked in a ratatouille. Tomatillos are rich in vitamins (A, B, C, D, K), improve digestion, and have diuretic properties.

Like tomatoes and Cape gooseberries, tomatillos thrive in warm, sunny environments, which requires growing them under cover, in a greenhouse or tunnel, after being sown in a greenhouse or warm, bright room as early as March. These rather large plants should be transplanted with a 24- to 28-inch (60–70 cm) spacing in all directions, and their bushy, sometimes unstable foliage may require trellising.

In terms of maintenance, plants require little care, aside from regular watering in summer while fruit is developing.

Harvest tomatillos from August to late October. Note that tomatillo flowers are self-sterile, meaning you must grow at least 2 or 3 plants to generate fruit.

Eggplants

Cultivated since antiquity in India and China, eggplant made its way to Europe in the 14th century, first in Spain and then eventually in France during the 17th century. This annual species of the Solanaceae family grows to 24 to 30 inches (60–80 cm) tall. The bushy plants have oval pale-green leaves, downy sometimes purplish stems, and a thorny calyx. Flowers have white or purple corollas (petals) and resemble potato blooms. Fruits are fleshy, smooth, and elongated or round and either yellow, purple, white, or black. Eggplant is harvested when it is nearly ripe and should be eaten soon after that. Fruits are rich in vitamins (A, B1, B2, B3, B5, C), calcium, chlorine, copper, iron, manganese, magnesium, phosphorus, sodium, and sulfur, making them a nutritious food with many health benefits. Baked, stuffed, cooked au gratin, fried into fritters, or eaten as an appetizer spread, eggplant pairs perfectly with rice and is a key ingredient in ratatouille and moussaka. Its many different uses make it an essential summer vegetable.

Eggplants

COMMON NAMES: Eggplant, aubergine.
SCIENTIFIC NAME: *Solanum melongena.*
FAMILY: Solanaceae.
REQUIREMENTS: Eggplant is a heat-loving crop that grows well in sandy loam soils in tunnels. Field cultivation is possible in a Mediterranean climate.
SPACING: 1 row down the middle of the bed (30 in. or 75 cm bed width), set 18 inches (45 cm) apart.
SEEDING: Professional growers can keep trays on heating mats from late February to April. Home gardeners can use a bright heated room, such as a sunroom.
TRANSPLANTING: In May after the last frosts.
DAYS TO MATURITY: 150 days.
ENEMIES: Colorado potato beetles, whiteflies, aphids, thrips, powdery mildew, botrytis.
VARIETIES: Monstrueuse de New York (round), Longue Violette Hâtive (Asian type, very early), Black Beauty (large classic eggplant), Annina (purple and white), Aubergine de Barbentane (long fruit), Rotonda Bianca Sfumata di Rosa (small, round white fruits with a purple blush), Clara (oval, white), Blanche Ronde à Œuf (dwarf variety, ideal for pots), Violette de Toulouse (purple, club-shaped).
A NOTE FROM JEAN-MARTIN FORTIER: Eggplant is absolutely worth growing in a vegetable garden. It is just as productive as cucumbers and sweet peppers, requires good bed preparation, and needs careful pest monitoring.

Growing Eggplants

"With shades ranging from violet to deep purple, eggplant has visual appeal and food lovers can't get enough of it."
Jean-Martin Fortier

Planting

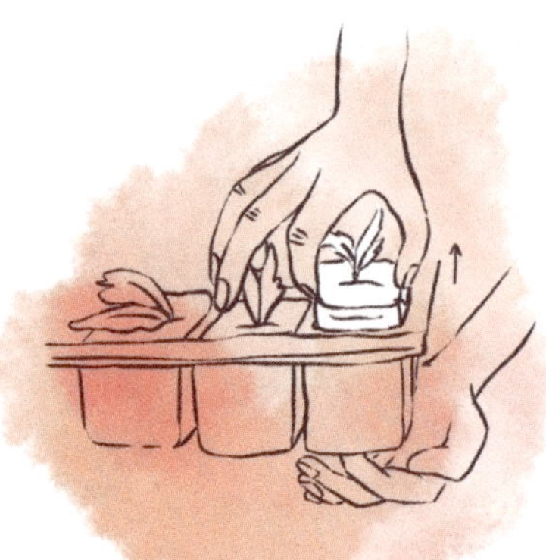

Seeding Under Cover

Professionals can sow eggplant in a greenhouse, while home gardeners can use a mini greenhouse or bright heated room. Seven to 8 weeks before the last frosts, sow into plug flats filled with moist potting mix. Cover with a thin layer of the mix and water the surface with a gentle spray. You can also sow eggplant in open flats. Keep the soil temperature between 81°F and 86°F (27–30°C) until germination. Once they sprout, maintain the space at roughly 75°F (24°C) during the day and 64°F (18°C) at night, making sure the soil is always moist.

Potting Up

Three weeks after sowing, transplant seedlings into 4-inch (10 cm) pots filled with potting mix. Put them in a bright space kept around 79°F (26°C). Water regularly.

Tip from Jean-Martin Fortier

About 4 to 5 days before transplanting seedlings, harden them off by exposing them to outdoor temperatures and reducing watering. Around this time, add a pinch of magnesium to every pot.

Preparing the Soil

In a greenhouse, or outdoors in more southern regions, loosen the soil with a broadfork then spread a 2-inch (5 cm) layer of chicken manure and compost. Run a power harrow set to a 2-inch (5 cm) depth down the bed to loosen and level the surface to promote good root development and warm the soil.

2

Professional growers can set up four lines of drip tape secured by fabric staples and covered with a woven ground cover or a tarp that has 4-inch (10 cm) holes every 18 inches (45 cm), prepared by a cutting or burning tool for mulches.

Tip from Jean-Martin Fortier

Of all the vegetable crops, eggplant is the heaviest feeder. To prepare beds before laying a woven ground cover, we recommend amending the soil only once, so the input must have significant nutrients: chicken manure, alfalfa meal, magnesium, potassium.

Transplanting

Transplanting into a Greenhouse

Before transplanting seedlings, make sure the soil is at least 64°F (18°C), ideally 68°F (20°C). Lay down a woven ground cover with holes (same diameter as pots), made using a blowtorch or hole burning tool, to make planting easier. Carefully remove seedlings from the pots then use a dibber to gently bury the root ball level with the surface.

2

Use a sledgehammer to drive 4-foot (1.2 m) metal stakes 12 inches (30 cm) into the ground, keeping 3 plants (so 5 feet or 1.5 meters) in between each one. This will be your trellis structure. Keep the soil moist for 1 week after transplanting eggplants.

Tip from Jean-Martin Fortier

The key to successfully transplanting eggplants lies in thoroughly watering the seedlings and the ground before planting to help the roots get established quickly. This also prevents air pockets from forming around the roots.

Transplanting into a Field or Garden

1 Transplant seedlings into the field or garden space in premoistened soil spaced 24 inches (60 cm) apart. Use a dibber to dig holes the same width as the pots but a little deeper. Add a handful of both compost and organic fertilizer and incorporate into the soil. Plant seedlings, tamp the soil, and water thoroughly.

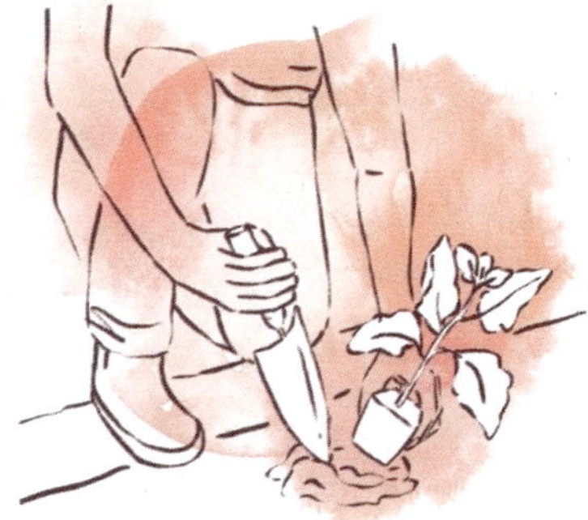

2 Cover with a 4-inch (10 cm) straw or flax straw mulch and water the base of each plant to settle the mulch and soak the soil. Watch out for slugs and snails that can significantly damage small seedlings. Early in their growing stage, use cloches or low tunnels to protect plants from the cold if late frosts are forecasted.

Tip from Jean-Martin Fortier

Dwarf varieties can be grown in 5-gallon (20 L) pots in a sunny location: a terrace, balcony, or courtyard. Put a layer of clay pebbles in the bottom of the container, which should have holes, and add potting soil mixed with 2 handfuls of organic fertilizer (horn meal). Cover with a layer of straw mulch to keep the soil moist. Eggplant cultivars, such as Ophelia, Blanche Ronde à Œuf, Rotonda Bianca, and Modern Midget, are particularly well suited to container gardening.

Maintenance

Trellising

In a greenhouse, weave nylon lines between metal stakes once plants are 16 inches (40 cm) tall, a technique known as the Florida weave. As they grow throughout the season, add 1 or 2 tiers of nylon line to support the stems, usually 1 tier for every 12 inches (30 cm) of growth.

In the field, drive 4-foot (1.2 m) stakes into the ground, 12 inches (30 cm) deep, right after transplanting to keep seedlings out of the aisles. Tie the same nylon line to the main stem, loosely to avoid strangling the plant.

Weeding

In a greenhouse or tunnel, hoe the soil and pull any remaining weeds by hand. Repeat the operation 2 or 3 times throughout the season.

Tip from Jean-Martin Fortier

If you choose to mulch the soil with straw, in the greenhouse or field, you must remove the mulch before weeding. Use a stirrup hoe or garden claw to both weed and loosen the soil. When finished, replace the mulch over the surface.

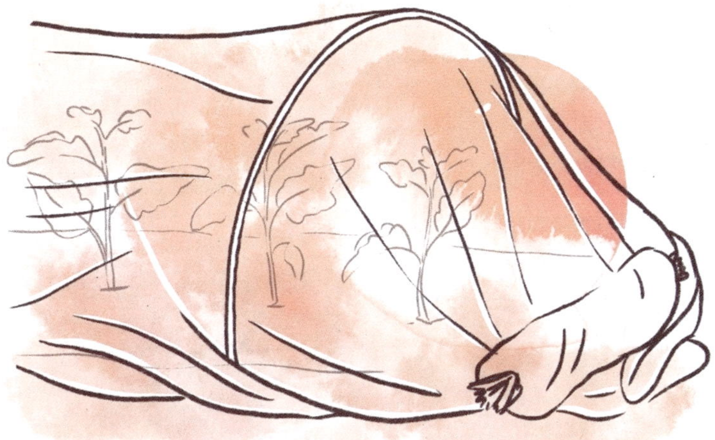

Plant Protection

Monitor each plant for potato beetles once a week as this mighty foe quickly devours leaves and decimates a crop. Since they are easy to spot, you can pick off and eliminate them by hand. For a heavy infestation, we recommend using a biopesticide.

In hot or dry periods, leaves may turn gray and slimy, and a black mold may dry them out, a sign of a whitefly infestation. Apply a biopesticide as soon as possible to keep it under control.

To prevent late blight, keep the foliage dry use a drip irrigation system. Otherwise, direct water towards the base of each plant and water with a nettle-based fertilizer tea a few times. If early signs of blight appear, treat plants with 2 or 3 applications of Bordeaux mixture fungicide.

Tip From Jean-Martin Fortier

To protect crops from Colorado potato beetles, lay insect netting over hoops immediately after planting outdoors. This combined with mulching protects the crop from wind and pests, providing a considerable boost for seedling growth.

Pruning

Professional growers do not need to prune the foliage; simply remove damaged leaves and suckers growing from the bottom of the main stem.

We recommend home gardeners prune to speed up and promote consistent fruit development in the field or in a greenhouse.

1. **Cut the main stem just above the first flower to encourage development of several side shoots. Remove suckers from the base to direct energy towards developing fruits.**

2. **Pinch off the tips of stems that grew back after the first pruning. Starting from the main axis and for each branch, count 2 or 3 flowers and cut the stem above a leaf. This will yield 6 to 12 eggplants per plant along 3 or 4 stems.**

Harvest

Picking

Twice a week from mid-July to October, use a harvest bag to collect eggplants of the preferred size, cutting the stem with curved or straight pruning shears. They should be firm and have fully developed color.
Hold fruits in one hand and cut the stem about 1 inch (2 cm) long. Gently place eggplants in a crate or harvest bin.

Storage

There's no need to wash eggplants, just wipe off any dirt with a cloth. They can be harvested at any time of day but must quickly be stored in a room set to 54°F (12°C) or the bottom drawer of your refrigerator to stay firm. They should be eaten fairly soon after harvest.

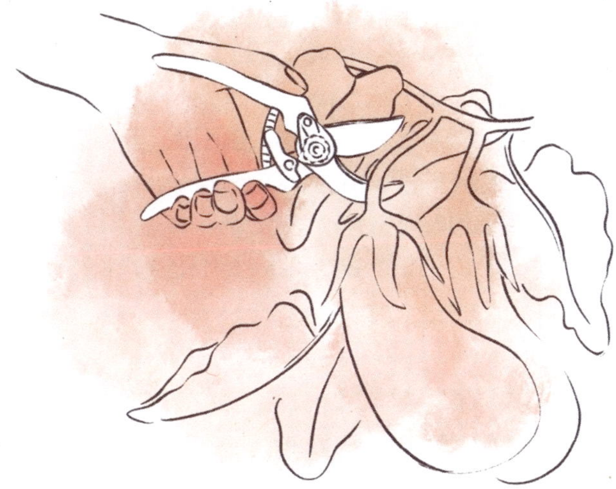

Tip from Jean-Martin Fortier

As with tomatoes, you can dehydrate thinly sliced eggplants: leave them in the sun for a few days, use a dehydrator, or use your oven on a low temperature. They can then be vacuum-packed or stored in airtight containers.

Sweet Peppers

Sweet peppers, native to South America, were brought to Europe likely in the 16th century. This annual typically forms a 30-inch (80 cm) bush with shiny, pointed green lanceolate leaves. White flowers bloom in late spring and develop into fruits that are round or conical, sometimes elongated, with a smooth, shiny skin. Fruits can weigh up to 11 ounces (300 g) and are red, yellow, green, or purplish-brown, depending on the variety and ripeness. The thin flesh is watery and slightly acidic but becomes more mellow, even sweet, once ripe. These vegetables have one of the highest levels of vitamin C. They are also rich in minerals (copper, iron, magnesium, manganese, phosphorus, potassium, and zinc) and has digestive, diaphoretic, and stimulant properties. They are easier to digest when peeled and can be eaten raw, finely chopped in salads, appetizers, or condiments. Steamed, baked, braised, peppers can be stuffed with meat and used in ratatouilles.

Sweet Peppers

COMMON NAME: Sweet pepper, bell pepper, pepper.
SCIENTIFIC NAME: *Capsicum annuum.*
FAMILY: Solanaceae.
REQUIREMENTS: This Mediterranean vegetable is rather sensitive to cool weather and must be grown under cover, in full sun, and sheltered from the wind. It thrives in cool, deep, humus-rich soils that warm up quickly in the spring.
SPACING: 1 row per bed or 2 set 10 to 12 inches (25–30 cm) apart, with in-row spacing of 12 inches (30 cm).
SEEDING: From February to March; in a greenhouse, for professional gardeners; in a mini-greenhouse or seeding flat in a bright heated room, such as a sunroom, for home gardeners.
TRANSPLANTING: In May, after the last frosts, preferably in tunnels if growing in cooler climates.
DAYS TO MATURITY: 120 to 130 days.
ENEMIES: Sweet peppers are generally quite vigorous and have few enemies. However, watch for aphids, European corn borers, and spider mites in tunnels and check for black rot on fruits and late blight on seedlings.
VARIETIES: Corne de Taureau Rouge, Chocolat (brown fruit, mild flavor), Très Long des Landes (early, sweet flavor), Calwonder Golden (yellow fruit, mild flavor), Sweet Banana (yellow fruit turning red at maturity).
A NOTE FROM JEAN-MARTIN FORTIER: Tricky to grow, sweet peppers need attention and care. But this sun-soaked vegetable is a must-have garden crop and an integral member of the summer veggie quartet, along with zucchinis, eggplants, and tomatoes.

Growing Sweet Peppers

"Sweet peppers love heat and develop flavor as they slowly ripen on the plant. They can even become quite sweet!"

Jean-Martin Fortier

Planting

Seeding

Home gardeners can sow sweet peppers from February to March in an open flat or a 3.5-inch (8–9 cm) pot in a warm bright space. Market gardeners can sow into plug flats kept in a greenhouse, on heating mats to maintain soil temperature at about 81°F (27°C) until germination.

In both cases, seed at a depth of about 0.25 to 0.5 inches (0.5–1 cm) then cover with a thin layer of soil, firm down, and water with a gentle spray. Keep the soil moist until germination. Air temperatures should be 64°F (18°C) at night and 75°F (24°C) during the day.

Potting Up

Once the seedlings have 2 to 4 healthy leaves, about 3 weeks after sowing, transplant them into 4-inch (10 cm) pots. Dig a hole in the potting mix a little wider than the root ball. Gently pull out and insert the seedlings, lightly tamp, and water with a gentle spray.

Transplanting

South of the 45th parallel, sweet peppers can be grown in the field. Further north, it's best to grow them in a greenhouse or tunnel. Home gardeners can plant sweet peppers under a plastic tunnel where temperatures are kept above 59°F (15°C).

Cover the soil with organic mulch. Professionals should cover them in greenhouses with a recycled woven mulch.

1. **Once plants are 6 to 8 inches (15–20 cm) tall and have their first flower buds, place them outside for 4 to 5 days under a floating row cover and water less frequently.**

2. **Loosen the bed, add 2 inches (5 cm) of compost and fertilizer (alfalfa meal or powdered chicken manure), and incorporate it with a tilther set to a depth of 2 inches (5 cm). Mark a single row with a taut string down the bed or use a row marker tool to draw two rows.**

3. **Every 12 inches (30 cm), dig a hole by hand or with a dibber, and plant a seedling. Bury the root ball to just below the first row of leaves. Thoroughly water the base to settle the soil and keep the ground moist—but do not overwater.**

Tip from Jean-Martin Fortier

For market gardeners, woven ground covers are beneficial as they hinder weed growth and keep the soil cool. Use a blow torch or hole burning tool to make holes in the ground cover before planting.

Maintenance

Trellising

After planting sweet peppers in a greenhouse or tunnel, tie nylon lines on the overhead structure to hang over the seedlings and connect them with tomato clips.

For field-crops, insert a 4- to 5-foot (1.2–1.5 m) wooden stake beside each plant. Loosely tie the stem to the stake with cotton line every 8 to 12 inches (20–30 cm).

Pruning

Regularly prune the foliage after peppers form to promote better fruit development. When 2 or 3 healthy peppers form, trim the end of the stem beyond a leaf growing after the last fruit. At the end of the summer, once each plant has 12 to 15 fruits, top the main stem so its energy can focus on the remaining fruits and improve flavor.

Tip from Jean-Martin Fortier

The first time you prune, you can remove some fruits. If a stem has too many buds and fruits, remove 1 out of 2 by snapping the stem just above the fruit or flower between your thumb and index finger. Keep only 2 or 3 fruits per secondary stem. The excess young fruits, even the tiniest ones, can be eaten and are particularly tender!

Irrigation and Hoeing

If the beds are not mulched, hoe them to eliminate weeds, loosen the soil, and improve water infiltration. Water regularly during fruit formation, but not too much. A drip irrigation system is ideal when programmed to water every morning (2 hours after sunrise) and afternoon.

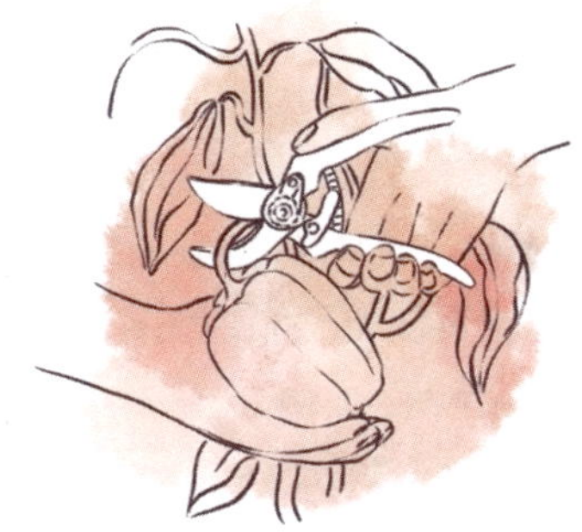

Plant Protection

To tackle an aphid infestation, cut off damaged leaves and introduce predatory insects such as ladybugs or lacewings under shelter and in a closed environment (tunnel, greenhouse). As a last resort, apply a pyrethrum solution (biopesticide). Pepper skins can be damaged by early blight (Alternaria); decay at the bottom of the fruit is blossom-end rot. As a preventive measure, ventilate greenhouses and tunnels, remove affected fruits, and water sparingly during the fruit development phase. For disease control, spray with a calcium chloride solution.

Harvest

Picking

Harvest from June to September about twice a week, when fruits reach the desired ripeness and color. Gently hold a fruit in one hand and cut the stem using pruning shears or a sharp knife. Leave less than an inch (1–2 cm) on the fruit.

Storage

Sweet peppers develop their full flavor when ripened on the plant, not in storage, so they should be eaten soon after they are harvested by professional and home gardeners. In a cold room set to 54°F (12°C), store sweet peppers for no more than 8 to 12 days. At room temperature, they will keep for 3 to 5 days.

Tip from Jean-Martin Fortier

Do not wash sweet peppers in water; simply wipe with a cloth to remove dirt and dust. This polishes the skin, giving them a glow that makes them even more appetizing and enticing on display.

Hot Peppers

Perennial but cultivated as annuals, hot peppers are native to Central America and were brought from Mexico to Europe in the late 15th century by Spanish explorers who called them cayenne peppers. Plants grow into a 24- to 32-inch (60–80 cm) bush with oblong green leaves. White flowers bloom in summer and form spherical hollow fruits, sometimes straight or curved, long or short, with a pointed tip. Their thin flesh can be yellow, orange, red, green, or purple, almost black. Rich in vitamin C, when fresh, and high in vitamin A, B6, B3, E, and K, fruits also contain iron and manganese and have diuretic and diaphoretic properties. Their high levels of capsaicin, an alkaloid, gives them a spicy taste that is measured on the Scoville scale, which indicates pungency (heat). The mildest peppers are at the bottom of the scale, while the hottest are at the top. They are eaten both fresh and dried, sometimes ground into a powder, and combined with other foods to offset the heat.

Hot Peppers

COMMON NAMES: Hot pepper, chili pepper, hot chili.
SCIENTIFIC NAME: *Capsicun annuum.*
FAMILY: Solanaceae.
REQUIREMENTS: Hot peppers are cold-sensitive and require a sunny, warm location sheltered from the wind. They thrive in loose, well-drained rich soil.
SPACING: 1 row down the middle of the bed or 2 staggered rows 14 inches (35 cm) apart, with plants every 12 inches (30 cm).
SEEDING: Market gardeners can sow in a greenhouse from February to March in open flats or plug flats on heating mats. Home gardeners can sow into pots or open flats in a mini-greenhouse.
TRANSPLANTING: In April in a greenhouse or tunnel. In May for field plantings after the last frosts.
DAYS TO MATURITY: 100 to 150 days.
ENEMIES: Although hot peppers can withstand pest damage, they may encounter some pests (aphids, cyclamen mite), diseases (anthracnose, scabs), and physiological conditions (blossom-end rot).
VARIETIES: Listed from mildest to hottest on the Scoville scale: Doux des Landes, Espelette, De Bresse, Rocotillo, Sucette de Provence, Cayenne, Pili Pili, Antillais.
A NOTE FROM JEAN-MARTIN FORTIER: Hot peppers can be distinguished from sweet peppers by their narrow elongated shape and, especially, much hotter taste. The capsaicin content differentiates hot pepper varieties and guides our choices.

Growing Hot Peppers

"Hot peppers are easy to grow and deserve a spot in every vegetable garden. Mild or strong, they set your meals and palate on fire!"

Jean-Martin Fortier

Transplanting

Transplant hot peppers in April into greenhouses and tunnels or from mid-May into the ground. Make sure to harden them off first. Plant seedlings every 12 to 16 inches (30–40 cm) in a row covered with straw or another mulch. Although their branches are strong enough to keep the bush upright, we recommend installing stakes and securing the main stem to them every 8 to 12 inches (20–30 cm).

Planting

Seeding

Hot pepper and sweet pepper crops are quite similar except that hot plants produce more fruits. Therefore the amount you sow must correspond to your desired yield. Starting in February or March, sow into open flats, plug flats, or pots in a warm, bright space (greenhouse, sunroom). Once 2 sets of leaves form, transplant into containers in a space at 68°F to 77°F (20–25°C) during the day and at least 59°F (15°C) overnight.

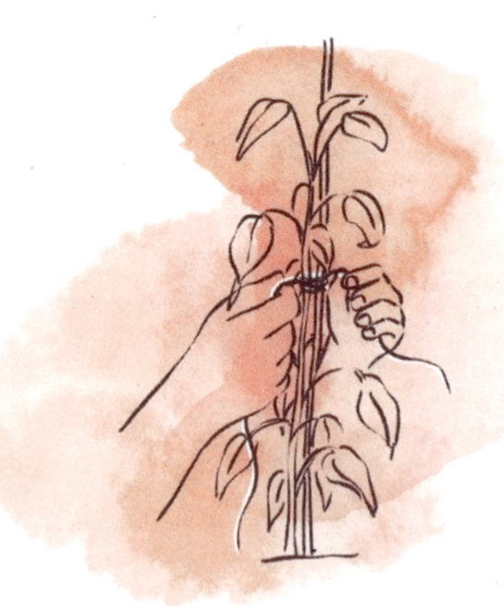

Maintenance

Irrigation
Although hot peppers thrive in hot environments, they also require consistent watering. Let the soil dry out between waterings and don't wet the leaves as this can foster disease. In addition to watering, apply a pelleted organic fertilizer.

Weeding
If the bed is not mulched, use a collinear hoe to loosen the soil and remove weeds as hot peppers are vulnerable to weed pressure.

Harvest

Picking
Hot peppers are harvested on demand from July to October after they develop a deep color. Use a knife or pruning shears leaving a section of stem attached to the fruit. Eat them fresh soon after harvesting as they keep for only a few days.

Storage
Leave a longer stem on Cayenne and Espelette peppers so you can string them up to dry in the open or in an oven at 122–140°F (50–60°C) or tied in bunches in a dry, well-ventilated room. They will keep for several years and can be eaten ground like pepper.

Acknowledgments from Jean-Martin Fortier

I wish to thank the entire team at the Market Gardener Institute for encouraging me to pursue my mission, every day. A big thank-you also goes out to the Growers & Co. team, who pushes me to come up with new types of equipment! I especially want to acknowledge my partner, Maude-Hélène Desroches, who is an exceptional market gardener and a dear friend!

Acknowledgments from New Society Publishers

We extend a great thanks to Delachaux et Niestlé, the French publisher, for working with us to publish this English edition. Further thanks to the New Society Publishers team for producing the book and especially to Laurie Bennett for her meticulous attention to technical details and high-quality translation into English.

Acknowledgments from Delachaux et Niestlé

A big thank-you to Jean-Martin Fortier and his team at the Market Gardener Institute for this wonderful collaboration.

Our heartfelt thanks go out to Pierre Nessmann for putting us in touch with Jean-Martin, for thoroughly editing this collection, and for being so generous with his time. For this book, we owe him so much. We also wish to thank Flore Avram, whose illustrations give this collection a beautiful, simple character; to Grégory Bricout for graphic design that cleverly reflects Jean-Martin Fortier's spirit; and to Sandrine Harbonnier and Sabine Kuentz for their work on the text.

Reference Books

The Market Gardener Masterclass, www.themarketgardener.com
The Market Gardener: A Successful Grower's Handbook for Small-Scale Organic Farming, New Society Publishers, 2014.
Winter Market Gardening: A Successful Grower's Handbook for Year-Round Harvests, New Society Publishers, 2023.
Microfarms: Organic Market Gardening on a Human Scale, New Society Publishers, 2024.

Grower's Guides from the Market Gardener

Tomatoes: A Grower's Guide
Vegetable Garden Tools: A Grower's Guide
Root Vegetables: A Grower's Guide
Living Soil: A Grower's Guide
Fruiting Vegetables: A Grower's Guide
The Well-Planned Vegetable Garden

Coming Soon

Fall and Winter Vegetables
Starting and Propagating Plants
Salads and Leafy Greens
Herbs
Perennial Vegetables

Translator: **Laurie Bennett**

About New Society Publishers

New Society Publishers is an activist, solutions-oriented publisher focused on publishing books to build a more just and sustainable future. Our books offer tips, tools, and insights from leading experts in a wide range of areas.

We're proud to hold to the highest environmental and social standards of any publisher in North America. When you buy New Society books, you are part of the solution!

At New Society Publishers, we care deeply about *what* we publish — but also about *how* we do business.

- This book is printed on **100% post-consumer recycled paper**, processed chlorine-free, with low-VOC vegetable-based inks (since 2002)
- Our corporate structure is an innovative employee shareholder agreement, so we're one-third employee-owned (since 2015)
- We've created a Statement of Ethics (2021). The intent of this Statement is to act as a framework to guide our actions and facilitate feedback for continuous improvement of our work
- We're carbon-neutral (since 2006)
- We're certified as a B Corporation (since 2016)
- We're Signatories to the UN's Sustainable Development Goals (SDG) Publishers Compact (2020–2030, the Decade of Action)

To download our full catalog, sign up for our quarterly newsletter, and to learn more about New Society Publishers, please visit newsociety.com.

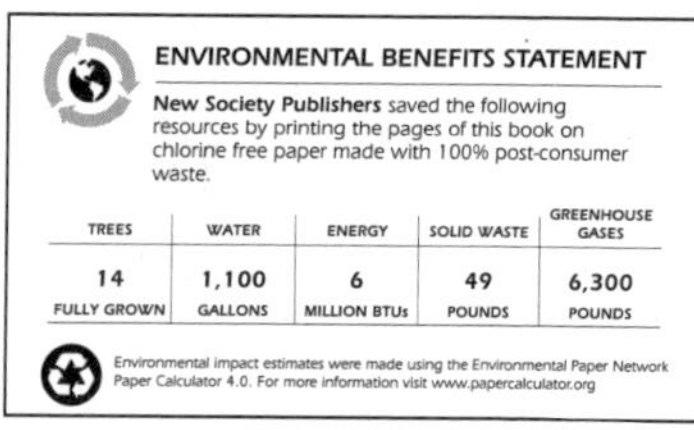

ENVIRONMENTAL BENEFITS STATEMENT

New Society Publishers saved the following resources by printing the pages of this book on chlorine free paper made with 100% post-consumer waste.

TREES	WATER	ENERGY	SOLID WASTE	GREENHOUSE GASES
14 FULLY GROWN	1,100 GALLONS	6 MILLION BTUs	49 POUNDS	6,300 POUNDS

Environmental impact estimates were made using the Environmental Paper Network Paper Calculator 4.0. For more information visit www.papercalculator.org